当代建筑师系列

崔愷
CUI KAI

崔愷工作室　编著

中国建筑工业出版社

图书在版编目(CIP)数据

崔愷/崔愷工作室编著.—北京：中国建筑工业出版社，2012.6
(当代建筑师系列)
ISBN 978-7-112-14305-4

Ⅰ.①崔… Ⅱ.①崔… Ⅲ.①建筑设计－作品集－中国－现代②建筑艺术－作品－评论－中国－现代　Ⅳ.TU206②TU-862

中国版本图书馆CIP数据核字(2012)第091506号

整体策划：陆新之
责任编辑：刘　丹　徐　冉
责任设计：董建平
责任校对：姜小莲　陈晶晶

感谢山东金晶科技股份有限公司大力支持

当代建筑师系列
崔愷
崔愷工作室　编著
*
中国建筑工业出版社出版、发行(北京西郊百万庄)
各地新华书店、建筑书店经销
北京嘉泰利德公司制版
北京顺诚彩色印刷有限公司印刷
*
开本：965×1270毫米　1/16　印张：11¼　字数：314千字
2012年8月第一版　2012年8月第一次印刷
定价：98.00元
ISBN 978-7-112-14305-4
　　　(22365)
版权所有　翻印必究
如有印装质量问题，可寄本社退换
(邮政编码 100037)

目 录   Contents

崔恺印象	4	Portrait
我是本土建筑师	8	I'm "Ben Tu / Native" Architect
中间建筑·艺术家工坊	12	The Inside-Out, Artist Colony
山东省广播电视中心	34	Shandong Broadcasting & Television Center
北川羌族自治县文化中心	50	Beichuan Cultural Center
泰山桃花峪游人服务中心	64	Tourist Service Center of Peachblossom Valley, Taishan Mountain
无锡鸿山遗址博物馆	78	Wuxi Hongshan Ruins Museum
前门23号	90	Legation Quarter
神华集团办公楼改扩建	102	Reconstruction of CSEC Office Building
浙江大学紫金港校区农生组团	112	Agriculture & Ecology Group of Zijingang Campus, Zhejiang University
南京艺术学院设计学院改扩建及南校门广场	120	Rebuilding of Design College, Nanjing Arts Institute
南京艺术学院图书馆	126	Library of Nanjing Arts Institute
北京外国语大学教学办公楼	136	Office Building of Beijing Foreign Studies University
北京外国语大学综合体育馆	148	Beijing Foreign Studies University Complex Gymnasium
承德城市规划展览馆	156	Chengde City Planning Exhibition Hall
崔恺访谈	166	Interview
作品年表	170	Chronology of Works
崔恺简介	180	Profile

崔愷印象

文／黄元炤

崔愷，1957 年出生，1984 年毕业于天津大学建筑系后，直接进入建设部建筑设计院，1985 年后被派往深圳与香港两地工作，1989 年调回北京本部，至此在部院（中国建筑设计研究院（集团））工作至今。崔愷是由国内建筑高校培养又选择留在国内发展的建筑师，尔后在部院内又成立了自己的工作室，所以，从 20 世纪 80 年代中期至今，在长达近 30 年工作生涯中，崔愷积累了厚实的实务经验，产生许多代表性且有意义的作品，也逐渐梳理与塑造出代表个人的建筑哲学观与中心思想。

崔愷于 20 世纪 90 年代后期有一件重要作品诞生，即北京外国语教学与研究出版社，又称为外研社。这个项目体现的符号暗示性语言尤为强烈，形式上为书与书架的意象，很大也很直接，看到的是建筑内外结合的一种韵味，这是崔愷当时一贯性使用的设计策略。其实崔愷更早期的西安阿房宫凯悦酒店就出现过相似的手法，官帽屋顶的暗示，是一种抽象的隐喻，在当时非常罕见。所以，崔愷在满足功能的基本需求后，设计倾向于一种符号、隐喻、象征与体量关系的语言，他将建筑视为几何体量的关系，个体的形是整体体量中的一个物件。而他塑造出来的体量关系，一方面回应了在那个时代背景下传统跟现代结合的期望，一方面体现出中国人对建筑的某种象征性的理解。更准确地说，崔愷的建筑欲表述出一种神似，崇尚某一种神韵，能更抽象点就抽象点，这种抽象化后的神似所散发出的韵味，是他个人的追求也是大众解读的需求。

外研社，由于是城市中的办公建筑，所以思考建筑跟城市的关系，在当时崔愷心中开始萌生。项目中巨大开口与开敞立面是面向城市而开放，仿佛特意与城市进行对话，这其实迥异于其他办公建筑所围合起来的私密性。而这样的开放性，一方面与功能和空间有关，一方面是以敞开胸怀的姿态，让城市看到内部空间层次的延伸与变化。但真正让崔愷更深入思考城市问题，是在日后通过国际交往与国外旅行时，他有机会在一个城市中停留且慢慢去观察城市，细细品味城市的空间。他对城市空间的兴趣，体现在日后的项目中，如德胜尚城办公小区。

崔愷，设计中首重功能性考量。所谓的功能性是透过对空间单元之间动态的排列组合后，展现出设计者所欲表述的设计行为与架构，探讨平面布局中物件与物件、单元与单元、列与列之间的前后组织相连的簇群关系。而崔愷认为功能是泛指的，除了以上所说的建筑内部使用功能外，功能还包括环境本身所提供的功能，室外功能对于人活动的引导与创造，城市功能与建筑功能之间的联系和对话等。所以，他认为功能是泛指的，这偏向于广义的功能定性，而功能考量后产生出神似的形体则是一个相对的结果，非设计意向上的目标。

崔愷，在偏向于广义性、泛指的功能基础上，设计姿态慢慢在转变，变得比以前更为开放。从阿房宫到丰泽园再到外研社，明显是一个强烈的形体隐喻与展现，着重于单体建筑的构成；到了北京外研社大兴会议中心，他有了改变，稍稍产生一种局部开放性，回归到功能与理性的思考范畴，注重群体功能的合理配置，平面布局依功能的增加而扩增，并逐渐有机地向外扩展，体现出单体与单体之间构成的群体前后组织相连的排列组合关系，此项目不再只关注单体，而是着重于群体建筑的构成，虽然姿态是开放，但反而收敛许多，看不到任何符号、隐喻与象征的语言，体量关系还是存在，朝向水平方向的发展。崔愷在这个项目中，思考到中国传统园林中的空间关系，也思考到后现代主义建筑中的景点制作，多样化链接的做法。

在姿态逐渐开放之余，思考建筑与城市的关系仍是当时崔愷的主要关注方向。他始终认为建筑是渺小的，建筑只是城市中的某个片段，他觉得传统城市空间比现代城市空间更有味道，更让他感兴趣。因尺度适合人们游走其中，可以慢慢散步去品味城市中隽永的人文情怀，所以在德胜尚城办公小区中，他关注到当地传统城市中文脉的历史延续，将原有胡同与院落的脉络，用现代的方式予以重现，考虑场地上传统城市遗留下的肌理与环境、与新建筑之间的融合，在功能的基础上，用肌理中的场景与线索将传统与现代混合、并置在一起。这样的场景与线索全来自崔愷儿时生活在胡同里的回忆，就是他自己的经验。在北京一定先从大街进胡同、进门楼后进院子，然后再进家里面，胡同空间的秩序变化是界定北京的一个很重要原则，门是朝里开，也是胡同中重要的一点，也因为这样使崔愷对文化的载体有了新的认识，重新去思考胡同——一个在北京很重要的建立社区文化的场所。

崔愷，从北京外研社大兴会议中心到德胜尚城办公小区，到之后的一些项目，不管是从场地、从园林、从城市、从文化的不同思考方向切入，在姿态开放后，他的形式语言渐趋于干净与简单，逐渐体现出一种几何体量的单纯化，去除繁杂的元素，这种去除也是一种"减"与"简"的态度，在形式与形体上减去后，能量不但不减，反而更强大，同时他的设计常常悠游于收放之间：河南安阳殷墟博物馆，从一个"洹"字，衍生出建筑形态，并结合国家对遗址保护，对地域景观的考虑，建筑低调隐没于土地之中，挖去的几何体与土地合一，而土地之上的建筑则归于零。北京韩美林艺术馆，建筑几何体退隐到巨大框架之内，低调且不张扬，是一种收的状态，然后才是清水混凝土肌理的表述，整体上内敛与质朴之美油然而生。山东广电中心，几何体量的语言更是明显，在基地狭长的限制下，东边以几何体与玻璃体相互穿插组成，与西边的高耸体量形成强烈的对比，夸张悬挑的体量，给人一种巨大的震撼力，建筑非常直接与开放。北京西山艺术工坊，在外立面上以不同进深凸凹的几何方块面与不同开窗面搭配，处理得更为干净与纯粹。

20世纪90年代后期，崔愷国际交流开始变得频繁。由于他作为一名中国的本土建筑师代表，自然受到更多的期待与质疑，进而也刺激到他自己的进一步思考。于是他认识到中国与许多亚洲国家一样，都在进行现代化建设，在追向欧洲，且价值观趋于一致，可是在实现现代化过程中，许多亚洲建筑师的关注点却又在另外一个方向上，他们拿出的作品是非常乡土的，都在讲他们的地域建筑、私人住宅、小会所，旧建筑改造等，很少讲他们大的现代建筑，这样的发现让崔愷思考到，虽然中国建筑师表面很国际化，但思想上是不是落伍了？另外他也反思到，作为一个生长在北京的建筑师，自己该对这个城市负起一些责任，并该建立起自己的价值观。于是，崔愷将自己的反思并同他多年来的实践过程与成果归纳与梳理后，总结出他的建筑哲学观。他认为建筑是一项综合性服务事业，不像可以主观表达个人情感的纯艺术，它有客观的因子。于是他提出"本土设计"的理念，"本土设计"在广义性、泛指的功能基础上，有着对城市历史与自然环境尊重的含义与态度，且需多加考虑建筑的各种影响因子，也是地域性与场所精神的追求，反应现状、满足功能，且以一种持中性的思考与态度，体现出作品多元、多面貌的形态。

之后，崔愷又深化了"本土设计"的理念，他认为"本土设计"即以土地为本的理性主义，他自己在设计中讲究设计的理性思考，而这又是现代建筑发展中重要的学派与原则，而这样的理性主义，是一种思考与思维模式，讲道理，有逻辑，更多的是关心环境、关心人文方面，所以，"本土设计"是立足于土地的理性思维。另外，"本土设计"在今天大规模建设环境中，也包括原则、立场与使用的策略，他认为设计是在这个基础上的一个更精致化、更高的艺术追求。所以，崔愷对于"本土设计"的态度是从容且客观的，而"本土设计"就是立足于土地的理性思维，表述的姿态趋近于优雅与自在，这样的设计逻辑与思考，是崔愷自己的建筑哲学观与中心思想。

Portrait

By Huang Yuanshao

Mr. Cui Kai was born in 1957 and graduated from the Department of Architecture, Tianjin University in 1984. He began to work for the Architecture Design Institute, Ministry of Construction, right after graduation and was sent to Shenzhen and Hong Kong since 1985. In 1989, he was transferred back to headquarter and stay there ever since. Mr. Cui Kai is one of domestic colleges graduates who chose to stay in China for his career and then founded his own studio in the institute.These 30 years since 1980s has provided him abundant practical experiences, so that he could have quite some representative and classical works, and form his own architecture philosophy and ideas.

A very important work of Mr. Cui Kai was presented in late 1990s, namely the Office Building of Foreign Language Teaching & Research Press, or Office Building of FLTRP. This project has a strong symbolism expression, with a straightforward huge image of books and shelf, showing a charm of combination of exterior and interior of the building, a staple design startegy of Mr. Cui Kai. In fact, one can clearly see the same style in Mr. Cui's early work, the Epang Palace Hyatt Hotel in Xi'an City: the roof looks like an ancient official hat to form a realistic and abstract hint which is quite rare at that time. Basically, when Mr. Cui Kai's design has satisfied all the fundamental function requirements, he would like to have his own design language: a symbolic, metaphorical one that reflects the relationship of building volume. He considers the building as relationships between different geometrical volumes,and the individual outline is nothing but part of the whole volume. This relationship he found also met the expectation of combination between modernity and tradition at that time,and on the other hand this relationship is also a Chinese symbolic understanding of architectures. To be more specific, with his buildings, Mr. Cui Kai wants to show a kind of spiritual similarity or a certain charm.He wants it to be more abstractive, since the abstracted similarity has a special charm, which is his pursuit as well as the need for common people to understand.

As an office building in center of a city, the Office Building of FLTRP makes Mr. Cui Kai begin to think about the relationship between architecture and city. The giant opening and the open elevation facing the city seem to have dialogs with the city, quite different from other office buildings that fenced for privacy. To a certain extent, this opening is due to the function and space requirements,and on the other hand, a opening style can also show the extension and changes of the internal space of the architecture. What makes Mr. Cui Kai further consider the city problems is his experience gained from his observation of other cities when he had the chances to go abroad for academic communication and travelling. His increasing interest in urban spaces began to show in his later works, such as Desheng Uptown.

Mr. Cui Kai always considers the functionality as the top important factor in his design. Functionality is normally the cluster relationships between objects, units or buildings in a plane layout, which shows the design behaviour and structure of the architect through the dynamic permutation and combination of different space requirements. Yet Mr. Cui Kai thinks functionality is a general idea,and besids the up-mentioned internal function of a building, it also should include the function provided by the environment itself, activities of people due the outdoor functions, and the connection and dialogue between urban functions and architectures. This idea of general functionality of Mr. Cui Kai takes functionality as a general idea, while the spiritual similarity that comes after consideration on functionality is a related result instead of what comes for the design aim.

With this idea of a general functionality, the design mentality of Mr. Cui Kai gradually becomes more open than before. From the Epang Palace to Feng Ze Yuan, and then to the Office Building of FLTRP, an evolution to a strong metaphor and exhibition of objects is obvious, with emphasis on an independent building. He changes a little with a sense of partial opening when design the Daxing International Conference Center of FLTRP in Beijing. He returns to the category of functionality and rationlism, and pays more attention to the proper layout of the cluster functions. Together with the increase of functions, the plane layout stretches as well, expanding outwards gradually, showing a relationship of permutation and combination of these connected clusters formed by different units. No longer does this project care only about the unit independently, but the construction of the building complex, even though these buildings seem quite open,and the design is much more inward. Due to lack of symbols and metaphors, the relationship of sizes still exists and extends horizontally. In this project, Mr. Cui Kai takes reference from the space relationship of traditional Chinese gardens and the design manner of multiple chains and the landscaping expression in post-modern architecture.

Still the thinking of relationship between architectures and cities plays keynotes on Mr. Cui Kai's design. He thinks that architecture is small, only a small part of a city, and he is more interested in a traditional urban space than that of a modern city, because he thinks that in a traditional urban space, the size is suitable for people to wander in, and the distance is controlled for a better taste of the ever-lasting humanistic feelings of the city. That is why in project of Desheng Up-town, to continue the history tradition, Mr. Cui Kai reappears the structure of HUTONG and yards in a modern way. He takes use of what left by the ancient city in that area, and integrates them with environment and the new architecture. This fusion of tradition and modern with scene and clues from the leftovers coming from Cui's memory of childhood, when he lived in HUTONG. In Beijing, one always comes from the street to a HUTONG and then through a gateway to the yard and finally to the home. This space order is a very important rule to define the city of Beijing. At the same time, in traditional Hutong culture of Beijing, the door always opens inwards. All these have made Mr. Cui Kai to reconsider the carrier of culture and reconsider the Hutong culture, a very important location for community culture in Beijing.

From the International Conference Center of FLTRP to Desheng Up-town and to other later projects, no matter how his thoughts are integrated into the design, whether the ground, garden, city or culture, Mr. Cui Kai is adopting a cleaner and simpler design characteristic with his more open thoughts, a simplification of geometrical blocks, and to unburden of multifarious elements. This can also be taken as an attitude of simplicity. The decreasing is only on forms and sizes, and the energy itself get stronger instead. The architecture is weakened and strengthened at the same time, which is a status between inward and outward. In the project of Yin Site Museum in Anyang, Henan Province, the shape of architecture derivated from the Chinese character of "洹", with consideration on requirements of protection and reservation of landscape view, the architecture immerged under the ground, the size and shape integrated with ground, part that above the ground decreases to zero. In the project of Beijing Han Meilin Art Gallery, the geometry of architecture withdraws into the giant framework, low-pitched and reserved, and quite inward. The fair-faced concrete texture spontaneously generates a feeling of introverted and modest. In the project of Shandong Broadcasting & Television Center, the geometry blocks become even more obvious, due to the long and narrow base. The eastern part consists of solid volumes and transparent volumes, together with the uprising volumes of the western part, forming a strong feeling of geometry. The high-sounding overhang struture shocks people, presenting a direct and outward feeling. In the Plot B of the Inside-out, cubes stretch out of the elevation with different depth with different windows, the whole architecture looks very clean and pure.

From late 1990s, Mr. Cui Kai frequently took part in international academic communication. During his several speeches, he was asked quite some questions. Maybe as a native architect that comes from China, he is bearing much expectation and question, which also arose his own thinking. He found that many Asian countries, China included, were in the process of modernization, and trying to catch up with Europe. The sense of worth in these countries was getting consistent with European as well. But at the same time, many Asian architects were paying attention to another field. Their works were very local. They seldom talk about large modern architecture, but only on local architect, such as private house, small clubs, or modification of old buildings and so on. These discoveries reminded Mr. Cui Kai that even though architects from China look very. internationalized, they may have been dropped behind in design thoughts. He also concluded that being an architect born and living in Beijing, he holds some responsibilities to the city of Beijing, and should have his own sense of worth. So Mr. Cui Kai concluds his own architect philosophy after he concluded and shaped his thoughts on architect over years of experences. He thinks that architecture is a comprehensive service business. Unlike the pure art that can express one's subjective feelings, architecture is concerning with objectives, and all related elements must be considered. That is why he raised the idea of "Land-based Rationalism". Based on generalized functionality, the Land-based Rationalism consists of a respectful attitude to the history of the city and the environment, and consideration of all elements may affect the architect, as well as the pursue to regional and local spirit. To reflect the current situation and to satisfy the functions requires neutral thinking and attitude for multiple looks of works.

He then further developed his idea about Land-based Rationalism. He considers Land-based Rationalism as a land-oriented rationalism, and pays attention to rationalism of design in his works, which is also an important shool of thoughts and priciples of modern architecture. The rationalism is a mode of think and logic, which concerns more about environment and humanity. In a word, Land-based Rationalism is a logical thought based on land. Under the circumstance of large-scale construction nowadays, it also means the principle, position and strategy that are being used, based on which he thinks design is a more elegant pursue to art. In a word, Mr. Cui Kai is trying to keep easy and objective with Land-based Rationalism, which is a logical thought based on land, with a more elegant and free expression. This design logic and thought is the core of Mr. Cui Kai's architecture philosophy .

我是本土建筑师

文／崔愷

我是本土建筑师。之所以自称为"本土",不仅是因为没有海外留学的经历,更因为提出了一个"本土设计"的主张。两年多前我出版了自己第二本作品集,为了起个恰当的名字颇费心思。回顾自己的创作历程,用什么概念能够涵盖这许多不同的项目后面不同的构思?换句话说自己干了二十五六年,有没有悟出一点门道,有没有形成一个明确的价值观呢?记得十年前出第一本小册子时,取名叫做《工程报告》,自认为水平还不到,不敢妄称作品,更谈不上设计思想。而那之后又忙了几年,似乎有些想法了。每每开始设计心里有些底数了,不那么盲目和纠结了。比如碰到城市中的项目,我知道要考虑建筑与现有城市环境的协调,还要关注城市公共空间的改善;而在自然环境中设计,就要尽量保护环境,让建筑融入风景之中;若是历史遗址旁的项目,最关键是文化态度要端正,谦卑而低调就不容易犯错误。而在西部少数民族地区的项目就要认真学习当地的文化传统,尝试用他们熟悉的语言去创作属于他们的建筑。当然我们更多的时候要面对现实生活,解决当下问题。如何在设计中以新的建筑空间语言去反映新的生活状态也是最经常考虑的事儿。归纳整理这些不同的设计策略,实际上就提出来一个最基本的问题:建筑创作以什么为本?如果我们把上述设计策略的依据条件抽象概括为自然和人文环境资源,所有建筑都会与之发生特定的联系,就像建筑无论怎样总离不开土地一样,那么"以土为本"是不是就是一个设计的基本立场呢?于是就提出了"本土设计"的初步概念。

当然,一个新概念的提出要得到别人的认同并不容易。一般人第一反应"本土设计"是强调市场保护,有排外的情绪。的确这和当时的英文译名有关,"native"是本国人或当地人的意思,很容易让人理解为本国人设计。之后在一些国际交往中送书给同行朋友,他们也往往有这样的疑惑。于是我在后来的演讲中就避免用这个译名了,宁可用汉语拼音"Ben Tu"来代替。去年夏天好友朱剑飞先生带墨尔本大学的学生来单位访问,正值我在家养病,便请他到家中小坐,叙谈中我提到译名一事,请他帮忙,几天后他发来邮件提出几个译法,我从中选了"Land-based Rationalism"一词,译过来是"立足土地的理性主义",比较接近我所说的"本土设计"的意思,希望能解决这个一直让我纠结的误读问题。

还有一些人认为从这个概念的内容看没有什么新意,似乎和"地域主义"区别不大,甚至有人担心会回到"民族形式"的老路上去,表现出文化的不自信,可能会影响建筑的创新。的确,从20世纪五六十年代兴起的地域主义建筑思潮是对现代主义国际式建筑的批判和纠正,提倡建筑对地域文脉的传承和发展,而为了强调创新,避免保守和折中主义,又提出了"批判的地域主义"主张。我认为"本土设计"的基本立场与之是一致的。但是在全球化经济、技术发展的背景下,国际建筑的主流仍然延续着风格化,品牌化或时尚化的路径,地域主义的发展一直处于边缘的状态,对城市特色的保持和发展影响有限,换句话说城市特色在全球化的冲击下不断削弱的势头并未改变。尤其我国快速城镇化的进程中,千城一面已经是不争的事实,建筑平庸化、商业化、功利性的现象普遍存在,不仅一直受到学术界的批评,也引起社会的广泛关注和不满。在这种情况下,仅仅提倡地域主义建筑显然不能解决问题,甚至会有人认为只有到偏远的乡村才能设计地域性建筑,而在中心大城市则不用考虑地域问题,仍然把地域主义误解为是边缘性的命题。我以为这不仅是对相关理论缺乏认识的问题,也多少与其名称的局限性含义有关,总让人觉得那是一种地方性的建筑策略,而不是建筑普适性的原理。之所以本人提出"本土设计"的观念,就是想让大家重新回归以土为本做设计的基本立场,重新确立建筑和自然与人文环境之土的依存关系,环境不同则建筑不同,建筑不同则城市不同,城市特色才能得以产生和延续。

另外,在这里之所以提"本土设计",而不是"本土建筑",主要是强调设计的立场,而不是特指某类建筑是"本土"。实际上从立足本土的立场出发,可以设计出多元化的、丰富多彩的建筑。比如强调建筑与城市历史的特定关系,可以设计出文脉建筑;关注建筑与气候环境的关系,可以设计出生态节能建筑;希望建筑融入自然风景之中,可以设计出地景建筑;力图表现建筑与大众生活的关系，可以设计出公民建筑，另外在乡村或少数民族地区强调建筑与地方文化和乡土营造的关系，还可以是乡土建筑或民族建筑等等。总之，"本土设计"应该是一个开放式的体系，既不强调一种风格，也不特指某种形式，这也就区别于以往有关"社会主义内容，民族形式"的主张。

前年三月底在天大开了一个有关"本土设计"的研讨会，大家对此发表了不同的看法，多数人认为"本土设计"是一种态度或是立场，但是有人又认为它有点儿悲情，有对抗性，甚至有点儿狭隘，这显然是有些误解，也可能是由于翻译成"native"造成的。还有人认为这是一种不自信的表现，这倒是说得有些道理，当下建筑界乃至文化艺术界对自己文化创新不自信是一种比较普遍的焦虑，但是如何摆脱焦虑，重新找回自信，恰恰是要回归本土，立足本土去找到创新点。更有人认为CCTV大楼也算本土，理由是它迎合了决策者的野心和价值观，对此我的确不敢苟同，因为值得质疑的是那种价值观真能代表我们文化的自信吗？其实细想想恰恰是这种内心的不自信需要由外在的某些东西来掩饰吧！还有朋友玩起了拼字游戏，挺有趣：比如把"土"放到后面，叫"本设计土"，若把"土"当作保守，就是批评，若是把"土"当作深厚的资源，那也可以说设计深入其中吧。若把前后颠倒——"设计本土"，意思也不错，为本土而设计，还有点儿责任感呢！虽然大家说法不同，但依我看大家干的事儿还是有共同点，每个设计都会讲来自地形、地段环境的启发，都会找点儿文脉的要素，只是侧重点不同，手法不同，水平也不尽相同，更重要是因为个性不同，作品自然呈现多元化的走向，这其实就是"本土设计"应该倡导的，换句话说这些各不相同的作品都可以装进"本土设计"的大筐里，而装不进去的就是那些生搬硬套的、模仿抄袭的、标准范式的、缺乏创新的设计了，而这类东西的确很多。

去年养病三个月，静下心来又重新梳理"本土设计"的想法，归纳出几句话："本土设计"是以自然和人文环境资源的沃土为本，强调的是以土地为本的理性主义设计原则。它既是一种立场，坚定地探索有中国特色的社会主义新建筑；也是一种策略，立足本土的建筑创作将以其特色重新奠定中国建筑在国际上应有的地位；更是一种文化价值观，是"和谐"这一中国社会核心的文化理念在建筑中的具体体现。它需要的是一种本土文化的自觉，反对全球化背景下的文化趋同性；它提倡的是对人居环境的长久责任，反对急功近利；它主张的是立足本土文化的创新，反对固步自封；它追求的是保持和延续不同地域的建筑特色，反对千篇一律的雷同和平庸。"本土设计"的理论框架可以比喻为一棵大树，它植根于我们丰厚的自然和人文环境资源的沃土中，通过理性主义设计的主干，生发出多元化的建筑现象：可以是地域建筑、文脉建筑，也可以是生土建筑、生态建筑，还可以是地景建筑等等。这就不局限在形式和风格上，而鼓励从更多元的方向上创作与本土环境息息相关的本土建筑来。

前几天忽然得知王澍先生荣获2012年度普利兹克奖，真的为他，也为中国建筑界而高兴！记得前些年有记者问过我何时中国人能获这个大奖，我那时随意推测可能要15年到20年吧。其实也是想我们这代人可能没戏，各方面条件还不具备，下代人从教育到实践，到产生真正的国际影响，掐指算来也差不多这个数，看看日本的获奖建筑师们也差不多，甚至更长久些。所以王澍的获奖的确有些突然，大大超出了我的预想。不过看到评委对王澍创作理念和作品的高度评价，让我感到在当今全球化，建筑界追逐明星和时尚的风潮中，植根于本土的创作道路再一次得到认同和关注，的确令人鼓舞！正如评委所说，王澍的作品虽然有明确的本土性，但创作方向上又超越了传统和未来的二元论题，对国际建筑学的发展具有普遍意义。联想到去年清华大学李晓东教授在福建土楼村里设计的一座廊桥小学荣获国际阿卡汗大奖，还有台湾的谢英俊先生的灾后重建项目和香港中文大学"无止桥"慈善助学的系列创作活动都得到国内外专业界的赞誉和褒奖，再次有力地证明了"本土的即是国际的"这个辩证的道理。另外，坦率说，相比起这些建筑同行，自己的作品还远不够"土"，自己对"本土"的思考还远不够深刻。但自己已经想明白了，不管别人如何理解，我将在这片沃土上耕耘下去，毕竟过了知天命的年纪啦。

I'm "Ben Tu / Native" Architect

By Cui Kai

I am a "Ben Tu [1]/ native" architect. I said so not only because of the absence of overseas educational experience, but also because of my proposition for design in land-based terms. Two years ago I published the second collection, for which to get a proper name I went to quite some lengths. In retrospect to the course of my creation work, what concept can I draw out to bring together so many different projects with distinctive ideas behind each? Or to be more precise, I ponder in introspect the set of values I have concluded as a result after being in this industry for over 25 years. I remember ten years ago when my first booklet, "Construction Report", was published, it was named so for fear that I should not have attained enough aptitude to brag of anything that can be termed as design concept. In the following couple of years, I busied myself around and managed to form some ideas. I'd know where to set off with each designing, not blindly or confusedly. For instance, when dealing with urban projects, I pay as much attention to the architectural harmony with the existing urban environment, so as to the improvement of urban public space; on handling designs with ecological concern, I put in priority the protection of natural environment, taking care the buildings be mingled unobtrusively to the landscape. As for projects next to historic sites, a proper attitude toward the culture is the key point. To be modest and unassertive, and many a mistake can be avoided. Projects in the ethnic prefectures in West China require conscientious study of local culture and traditions, aiming to create buildings with local peoples' familiar architectural languages. For sure, more often than not, we need to face the reality, and solve present problems. And how to reflect new states of life by using new architectural lingua in design is still another matter most often considered. These varied design policies in effect set forth the same basic question, i.e. what is the foundation of architectural creation. If we generalize the preconditions of the aforementioned policies abstractly as natural and human environmental resources, with which all the buildings will have necessarily distinctive relations, such as none can be built without the support of land, and then the design of a specific building should be based on its locative land, and in the general line of things, on the nation as a whole. And thereon I preliminarily formed the concept of native design.

Of course, it's not easy for a new concept to get mass recognition. Most people's first response to native design is that it accentuates market protection. Indeed, the word "native" (translated from Chinese. It should be kept in mind that translations are at times unavoidably equivocal to certain extent) can somehow be associated with nativism, or something done or designed exclusively by local people. When I gave books with such translated expressions to peers and friends from other countries, they also tended to have this doubt. So I avoided using "native" in pursuant speeches, preferring "Ben Tu" as its the pronunciation in its Chinese name. Last summer, my friend Mr. Zhu Jianfei came to visit my institute with his students from Melbourne University. Since I was staying at home recovering from illness, I invited him to my place to have a get-together. As we chatted on, I stated my said bewilderment, and days later, he sent me an e-mail proposing several possible translations. I picked out among his suggestions "land-based rationalism", which I think was comparatively close to my idea of native design, hoping it would solve this haunting problem of miscomprehending.

There are still others who don't think that this concept is anything new, that it bears much in difference from regionalism, and some people are concerned it will backtrack to "national form", which demonstrates a diffidence from one's culture, and can affect the innovation of architecture. Truth be told, the architectural trend of regionalism, arising in 1950s and 1960s and upholding the inheritance and development of architecture in the regional context, is a critique against and an attempt at rerouting Modernism International Style. And in order to emphasize innovation and to prevent conservatism and eclecticism, a "critical regionalism" is further put forward. I think the basic stance of native design is consistent with it. However, with economic globalization and technological advances, the architectural world will unswervingly follow its main path boosting style, brand and fashion, while regionalism progresses marginally,which limites of its impact on the maintenance and development of urban disposition. In other words, under the forceful wave of globalization, a city's distinguishable disposition is gradually ebbing away, and the trend is unlikely to change hitherto. This is especially true in China. In the rapid process of urbanization, it has become too obvious a fact to miss that many cities take on a same look, with run-of-the-mill and commercialized constructions almost everywhere. It not only has been the subject of academic criticism, but has aroused widespread concern and social dissatisfaction. In this case, the problem seems insoluble by merely preaching architectural regionalism, and some even argue that for the fulfillment of regional style, architects must go by themselves to remote villages. Yet it's quite another story in big cities, where regionalism is still misunderstood as a marginalized proposition. I think it due more than to the relative lack of theoretical understanding, as it's related more or less to the limited meaning of the name as well, considering the fact that people, by the use of this name, are led to take it as a localized building policy rather than a universal principle. Thus told, my idea of Land-based Rationalism aims at introducing our work back to the land-based approach, and to the re-establishment of buildings' dependency on the nourishing soil of natural and human environment. Different environments comprise different buildings, and different buildings constitute different cities. As a consequence, a city's distinctive disposition is produced and continued.

What's more, I refer to "native design" instead of "native architecture" here, so as to emphasize the standpoint of design. Actually, starting from this land-based standpoint, architects can design many diversified buildings, such as culturally-situated buildings stressing specific relations between construction and city history; energy-saving buildings addressing concerns about construction and climate; landscape buildings working at blending construction into natural scenery; citizen buildings trying to present the relation between construction and people's life; and indigenous buildings stressing relations between construction and local culture in rural villages or ethnic prefectures, to name a few. In short, "Land-based Rationalism" should be an open system, not dwelling on one style or a specified form, and thus should be different from the previous assertion on "socialistic content, national form".

Early in May, 2010, a seminar on native design was held at Tianjin University. Participants expressed different opinions on this topic, and when most people reckoned it as an attitude or viewpoint, some argued that it was somewhat pessimistic, antagonistic, or even bit of insular. The latter were apparently misunderstandings, caused as I conceived, once again by the translation of "native". There were still others who took it as a sign of diffidence. Their arguments held some grounds, as diffidence on the innovative ability has become a kind of pervasive anxiety running through construction industry to the gamut of arts and culture fields. But how to get rid of this anxiety and retrieve confidence? I believe we need to pick up from where we left off by focusing on what's local/native to find innovative points. Some people claim the China Central Television (CCTV) Tower is native, because it appeals to decision-makers' ambitions and values. In this regard I have my reservations. I doubt those values are sufficient enough to represent our confidence in our culture. It should be closer to the truth that they're employed sketchily to gloss over the diffidence deep in heart! Some would play with the words, kind of interesting: one is "design, native[2]". If nation is deemed as conservatism, this phrase should be a critique, but if nation refers to solid and rich resources, the design will be in the same vain deep-rooted. The other is "design the nation[2]", whose meaning is not bad, given that it's a demonstration of responsibility to design for the nation! Although people have different opinions, what we do share something in common, that is, of each design, we talk about the inspiration from terrain and environment, and the makings of cultural context, only that we shall take different focal points, make different approaches, have varied competence, and what's most important, everyone has different personalities, so that works won't be the same. This is what Land-based Rationalism should promote and push forward, or in other words, these diversified works should all be included in the assortment of native design, while those wont' be that are inflexible and stereotyped designs, and clumsy copycats, though they are not small in number.

During the three months last year when recovering from illness, I made use of the time to settle down and brood over my ideas of Land-based Rationalism. In summary, Land-based Rationalism is a rational design principle, taking the natural and human environment resources as its base, and emphasizing the importance of its supporting land and surrounding backgrounds. It indicates both a stance to unremittingly explore the possibilities of new socialist buildings with Chinese characteristics, and a policy for land-based constructions with distinguished features to gain the rightful place in international architecture. Moreover, it reflects a cultural value, serving as the architectural embodiment of "harmony" which is the core cultural idea of Chinese society. It requires awareness to native culture, opposing the convergence of cultures under the background of globalization; it encourages sustainable habitation conditions, opposing pursuit after fast fame and immediate profits; it supports innovation based on native culture, refusing to be fettered by past glories; it seeks the maintenance and development of architecture with different regional features, opposing monotonous and prosaic repetition. The theoretical framework of native design can be compared to a tree, which takes its root in the fertile soil of natural and human environment resources, and which through the trunk of rationalistic design, sprouts out a variety of architectural phenomena: such as regionally or culturally-situated building, indigenous or ecological building, or landscape building, etc.Thus not limited in form and style, more designs are encouraged to make aversatile creation of native constructions closely combined to native environments.

A few days ago I learned that Mr. Wang Shu won the 2012 Pritzker Architecture Prize, and was really glad for him, and for the architectural field in China! I remember years ago when a reporter asked me when the Chinese could win this award, I assumed without much thinking that it might take 15 to 20 years. I didn't expect it could be achieved by my generation, as we're not fully conditioned, so to speak. And it should take so much time, or even longer, for the next generation to get adequate education and practice, and rise to the eminence of international influence. So is the case with Japanese architectural prize winners. Thus told, Mr. Wang Shu's unexpected success was beyond my assumption. The Pritzker jury has given a high appraisal to Wang Shu's creational concept and built works, which as to me, is a positive nod to the native-based course of architecture irrespective of the current trend of globalization and fashion-fitting. As the jury said in its citation, "Wang Shu's work is able to transcend the debate as to whether architecture should be anchored in tradition or should look only toward the future, producing an architecture that is timeless, deeply rooted in its context and yet universal." It reminds me of Professor Li Xiaodong from Tsinghua University, who won Aga Khan International Award last year for his work of a covered-bridge primary school within an earth building in Fujian Province, and Mr. Hsieh Ying-Chun's post-disaster reconstruction project and the series charity project "Bridge Too Far" held by the Chinese University of Hong Kong which both received high academic praises from home and abroad. These are all powerful proofs of the dialectical fact as to be "of the nation, of the world". Frankly speaking, compared to these peers, my works are still not adequately "native", and my meditation on what's native can go deeper. But whatever others may think or take, I know, at my age, that I'll continue with my practice on this fertile land.

[1] "Ben Tu" is the Chinese pronunciation of "本土", referring to anything locally-made, or created with an elaborate consideration of its origin and immediate surroundings, and out of full deference to different related norms and makings, including history, culture and tradition, to name a few.

[2] The word property of "native" is changed from adjective to noun, according to the change of word order, and so is the case with the next expression. The author illustrates here different quotations from others, trying to providing plural perspectives for readers' comprehensive understanding of his idea about what's native design.

中间建筑·艺术家工坊 北京
The Inside-Out, Artist Colony, Beijing
2007~2009

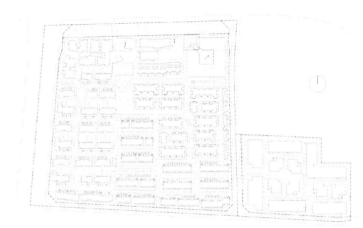

总平面图 / Site plan

当需要在一个108米×108米,限高15米的方形场地中安排75套艺术家工作室时,我们只能想到一个词——扎堆。有趣的是,这就是中国艺术家的生活状态,艺术家都喜欢扎堆。这是一次带策划色彩的艺术聚落设计,容纳艺术创作和生活的建筑单体通过群体规划的方式布置,相互毗邻,形成街道与大院空间;单体的聚集形成"圈子",底层是对大众开放的"艺术圈子";二层屋面上则是属于艺术家自己的内街社区——"生活圈子"。

我们用"漂浮容器"来形容单体中艺术家创作与生活的相互关系:工作就像潜水,一种憋足了劲的创作活动;而生活就像呼吸,一种悠闲的漂浮状态。这种沉浮状态在设计中表现为,工作空间处于底层5.4米的高大空间,而生活空间则像一个个白盒子悬浮其上,它们可以脱离,也可以进入下层工作空间,形成丰富的若即若离的空间体验。盒子里每层是7米×7米的方正空间,贯穿各层的功能墙的引入使得厨卫和管井等必备元素可独自形成一个功能岛,以强调居住空间的纯粹性和灵活性。

在整个建筑的西北角,艺术展示厅被提升到10米,让出一个模糊了内外界限的空间,打破封闭的"圈子",这里不仅是群落的主入口,也将是艺术家与大众互相碰撞和浸染的乐园。

摄影:张广源 Photo: Zhang Guangyuan

When the architects was asked to deposit 75 suits of artist studios in a 108m×108m site, with a height limitation of 18m, the only thing they can image is one word – stacking. What interesting, is that exactly describes the situation of Chinese artist's real life – artists are always gregarious animals. An art village, containing artists' creation and living activities, is formed here. Through the elaborate planning process, street and courtyard are encircle by a mass of modules, the assembly of individuals. On the first level is the opened "art circle" for public, while the second level is an inner street – "living circle" for artists' themselves.

The concept – "floating container" is employed to represent the interaction of working and living in the same studio unit. For artist, working looks like driving beneath water surface, holding his breath to do his best, while breathing freely well depicts his leisure life in the rest time. Living spaces are several pure white boxes, floating above the 5.4m high working spaces. Raised up or sunken down, they exhibit various spatial relationship between working and living. With a 7m×7m square planned for every floor, each white box is equipped with a functional island including all the necessary service pipes to maximize the flexibility of the living spaces.

On the northwest corner, a huge glass box, the exhibition hall for artists, is elevated to 10m, interrupting the enclosure of the "circle". Without legible boundaries of outside and inside, the intricate void serves as the main entrance of the art-studio cluster, and emphasizes the nature of this complex, a paradise of communication and infection for artists and public.

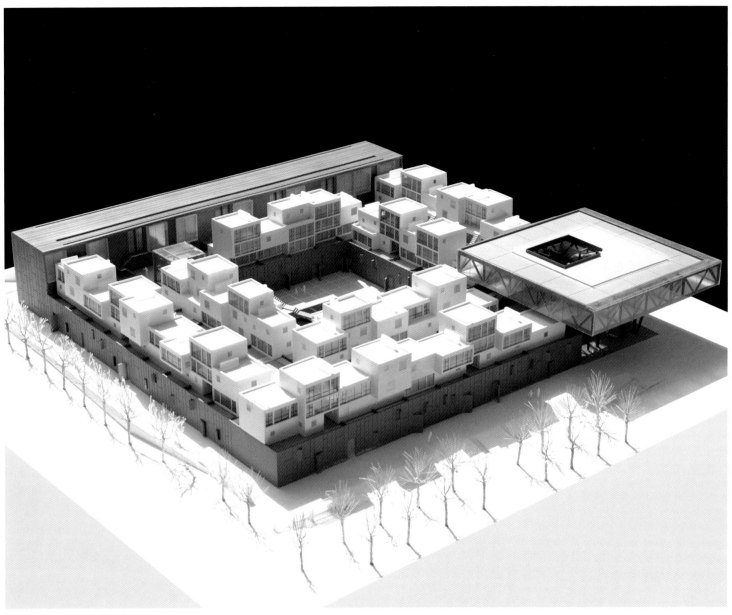

模型研究 / Model study

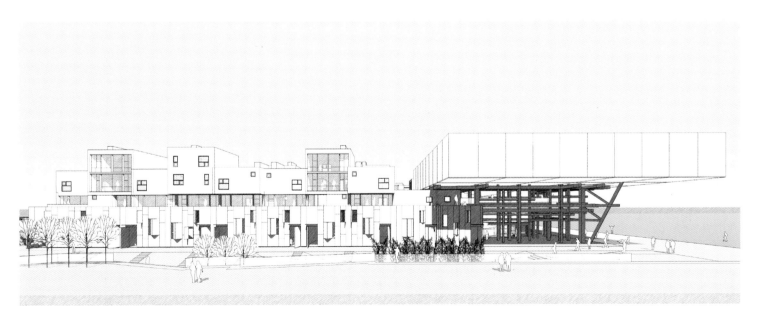

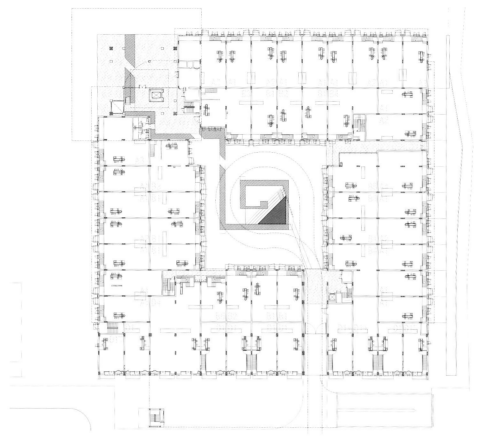

首层平面图 / The 1st floor plan

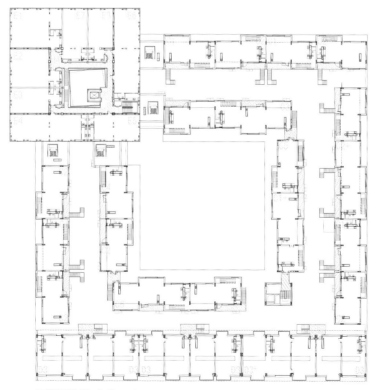

三层平面图 / The 3rd floor plan

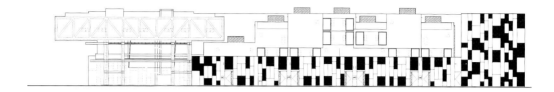

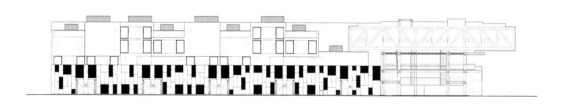

立面图 / Elevations

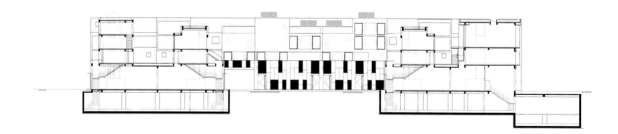
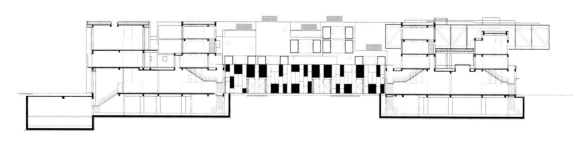

剖面图 / Sections

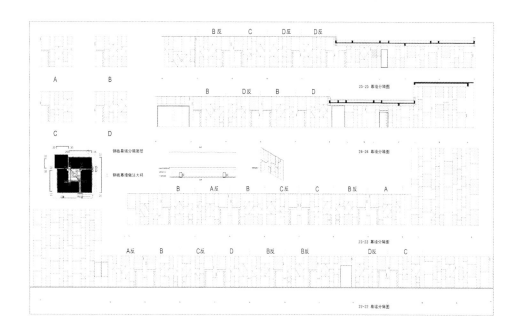

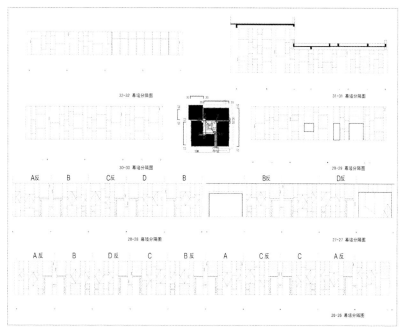

幕墙细部 / Curtain wall detail

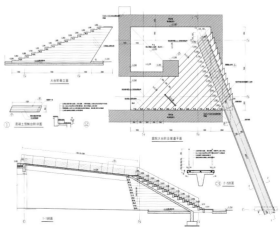

庭院台阶细部 / The yard steps details

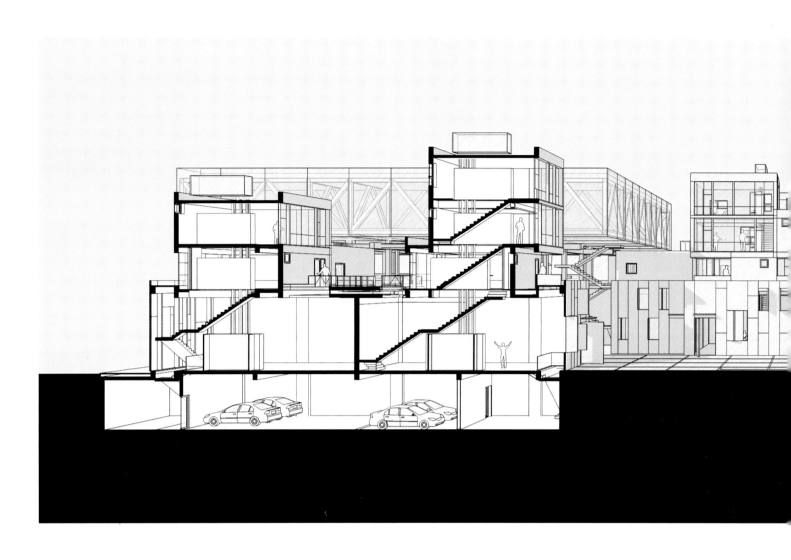

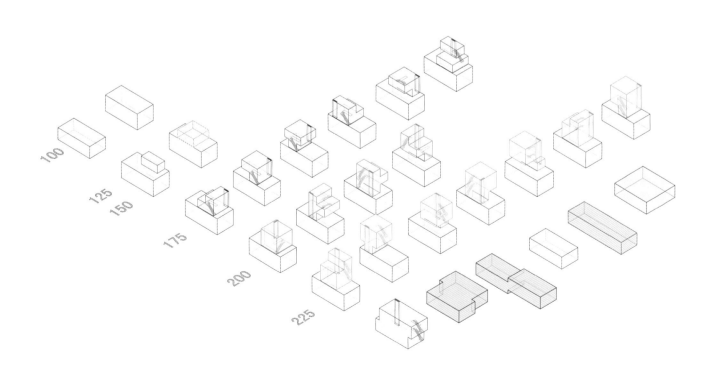

户型研究 / Forms of flat

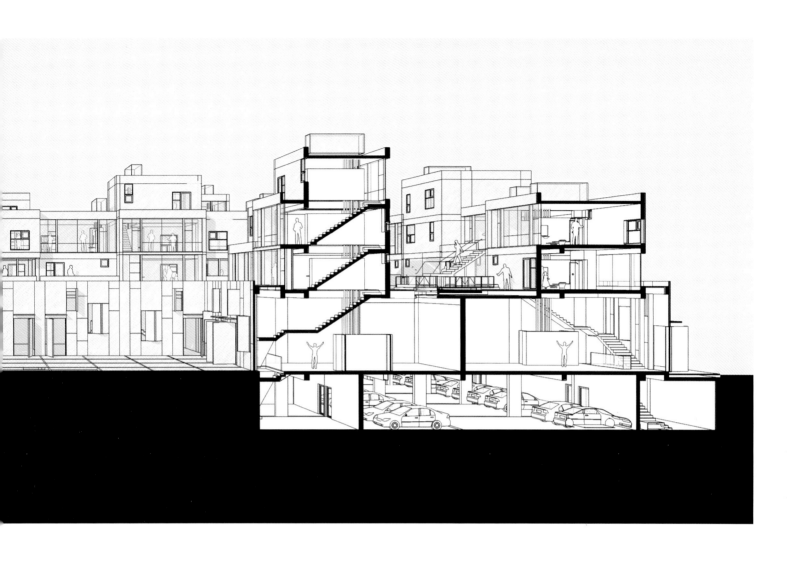

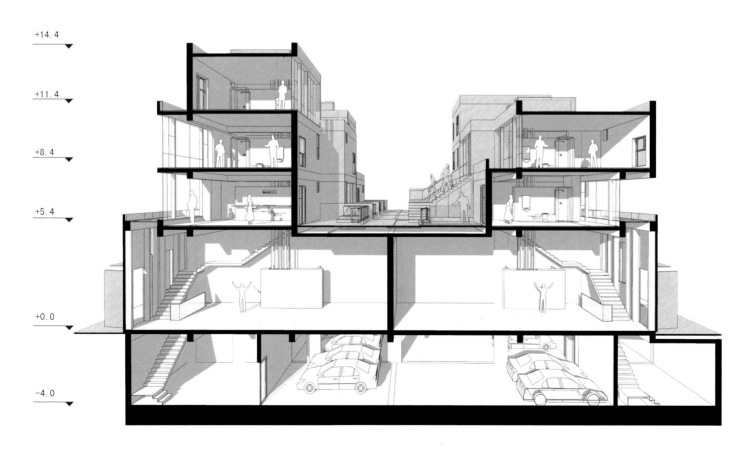

山东省广播电视中心　济南
Shandong Broadcasting & Television Center, Jinan
2004~2008

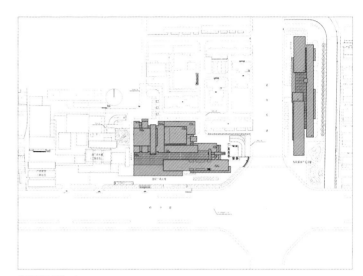

总平面图 / Site plan

　　山东省广播电视中心位于济南由千佛山至趵突泉和大明湖的景观轴线上。广电建筑普遍具有非常复杂的功能工艺要求，而作为原有山东省广电中心的扩建项目，且用地跨越主要干道，广电大厦的设计还需要解决与各个建筑互相衔接、统一改建的问题。

　　建筑群的西侧为新建的广电主楼和改扩建的旧建筑组成的广电中心综合业务楼，东侧为以各类经营活动为主的媒体中心。位于狭长用地上的建筑，通过连串排布的长条实体和玻璃厅虚实相间，空间形态明确而富有震撼力。建筑体量西高东低，在道路交叉口形成夸张的悬挑，使得建筑物无论对于行人，还是周边高层建筑的俯瞰，都保持了恰当的体形。两处建筑通过建筑语言、形态的整体设计，成为文化轴线上重要的节点。新建主楼西侧有原有主楼、技术裙房和广电宾馆等需要保留的原有建筑，通过对这些建筑原本掩藏的因功能不同而导致的不同层高进行充分暴露，获得了一种丰富而富有趣味的造型因素。

　　以"巨石"作为隐喻的外部形象，体现了山东的地域性和文化底蕴，同时也以巨型筒体结构支撑、悬浮和主楼层层叠合的形态，以及这一手法在室内设计中的延续，形成建筑强壮、简洁、富有力度的整体氛围。

摄影：张广源　Photo: Zhang Guangyuan

Occupied a spot on the landscape axis of Jinan city, from Qianfo Mountain to Daming Lake, the project was not only destined for accommodate those extra complicate functions, but also resolve all the existing problems of the various structures built in different stages. It should create a seamless, state-of-the-art facility.

A busy street splits the site into two parts: on the west site, a new-built tower dominates the comprehensive building as well as those reconstructed, while a new media center for various businesses is situated on the opposite side of the street. On both narrow sites, a series of solid slabs running horizontally and vertically, in contrast with the glass pavilions, define an intricate and impressive spatial system. The huge tubes are cantilevered extensively to gesture to the road, providing a remarkable image for pedestrians and views from the surrounding high-rise buildings. The expressive architectural language defines the site a pivot on the main axis of the city. Those existing structures are sufficiently exposed to emphasize the different story heights caused by the various functions, which also be treated as some interesting factors of television center.

Granite slates draw on the colors and textures of the mountains, reflecting natural environments of Shandong. Although not literal, the exterior image of the building was informed by the memory of huge mountains, an indigenous feature of Shandong. Accompanied by the interior design, the huge slabs and the exterior multi-layered composition reinforce the strong, solid and powerful image of the building.

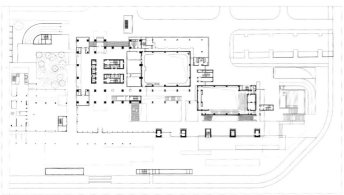

首层平面图 / The 1st floor plan of west building

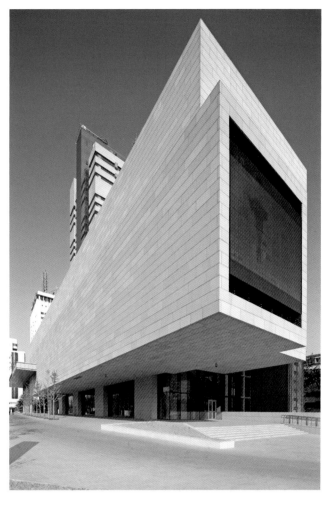

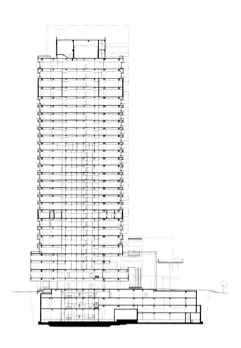

剖面图 / Section

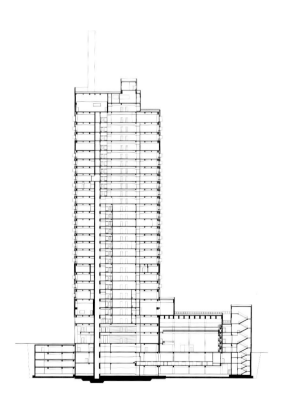

剖面图 / Section

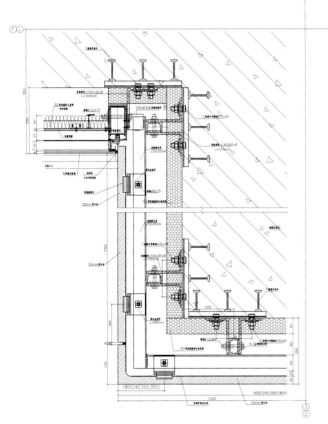

幕墙细部 / Curtain wall detail

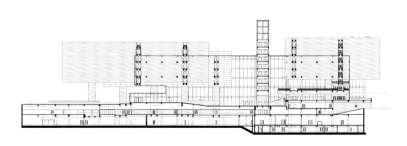

东院剖面图 / Section of east building

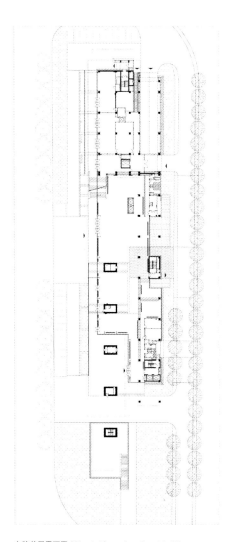

东院首层平面图 / The 1st floor plan of east building

北川羌族自治县文化中心　新北川县城
Beichuan Cultural Center, Beichuan
2009～2011

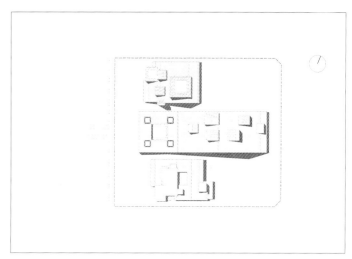

总平面图 / Site plan

北川文化中心位于新县城中轴线东北尽端，与抗震纪念园相邻，由图书馆、文化馆、羌族民俗博物馆三部分组成。设计构思源自羌寨聚落，以起伏的屋面强调建筑形态与山势的交融，建筑作为大地景观，自然地形成城市景观轴的有机组成部分，并与城市背景获得了巧妙的联系。开敞的前庭既连接三馆，也可作为各族人民交流聚会的城市客厅。建筑以大小高低各异的方楼作为基本构成元素，创造出宛如游历传统羌寨般丰富的空间体验。碉楼、坡顶、木架梁等羌族传统建筑元素经过重构组合，成为建筑内外空间组织的主题，并强调了与新功能和新技术的结合。

Situated in the northeast of the central axis of the new county seat, the Cultural Center is adjacent to the Earthquake Memorial Park, and composed of Library, Cultural Hall and Qiang Folk Museum. The design conception for the Cultural Center originates from the Qiang settlement, and emphasizes on the blending of the architectural form and mountains with fluctuant roofing; as the earth landscape, the building naturally forms an integrated component of the urban landscape axis, and subtly contacts with the urban background. The open forecourt not only connects with the three pavilions above-mentioned, but also can be considered as an urban hall in which people from all nationalities can convene and communicate together. In addition, taking square buildings of various sizes and heights as the basic element, the building complex creates an abundant space experience as if people are travelling in a Qiang village. Through reconstruction and combination, the Qiang traditional architectural elements such as turret, pitched roof and wooden beam become the theme of the building's internal and external spatial organization, which emphasize on the combination with new functions and technologies.

摄影：张广源　康凯　Photo：Zhang Guangyuan　Kang Kai

羌寨 / Qiang village

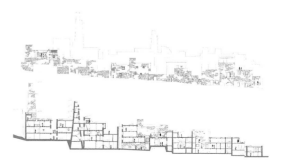

设计理念 / Design concept

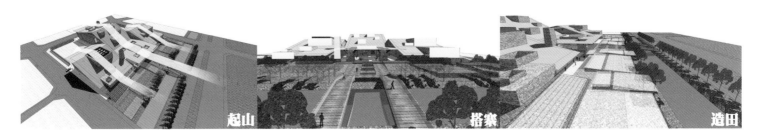

起山　　搭寨　　造田

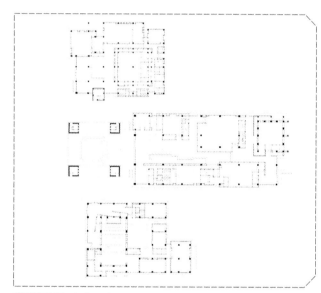

首层平面图 / The 1st floor plan

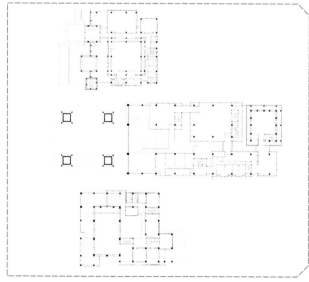

二层平面图 / The 2nd floor plan

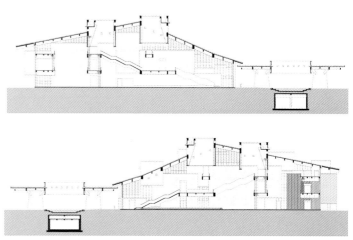

剖面图 / Sections

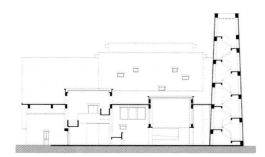

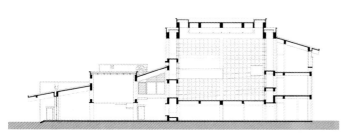

剖面图 / Sections

泰山桃花峪游人服务中心　　泰安
Tourist Service Center of Peachblossom Valley, Taishan Mountain, Tai'an
2008~2011

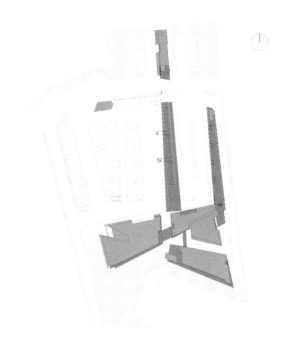

总平面图 / Site plan

　　桃花峪游人服务中心设于通往泰山的道路旁，南侧是由彩石溪汇流而成的南马套水库。设计保留基地内两处不同标高的停车场，并充分利用地形高差，用长长的坡道将上下山的游客流线以立体交叉的方式组织起来，互不干扰但能进行视线交流，满足游客中心人流的基本功能要求。湖水被引入建筑内部，并结合坡道设置，增加上下山的情趣。

　　接待大厅、餐厅、纪念品销售等主要建筑空间靠近湖面设置，充分利用优美的景色。在上山候车区设置候车廊和挑棚，面向泰山方向。办公、贵宾接待等放置在道路另一侧，建筑跨过马路形成关口，利于管理。

　　建筑形态充分呼应桃花峪独一无二的地貌特点，如彩石溪的石头一般，棱角分明，错落有致。当人们在"石头"之间的室外空间行走时，可以看到泰山雄伟的景象，建筑与自然山水融为一体。现浇彩色混凝土模拟彩石溪石头的肌理，形成朴素而自然的外观。

This project lies on the roadside to Taishan Mountain. To the south is a reservoir formed by stream from Color-stone River. Two park lots lying at different levels are reserved. And a long ramp is built to organize the flows of tourists coming and leaving, in which way the drop of the now existing topography was fully used. The two flow lines form vertical crossing, which make view contacts possible without interference to each other, which satisfies the fundamental function requirements of the tourist center in management of the flow lines. Water from the lake is introduced into the building, and the water channel goes along the ramps as a visual focus.

The reception hall, the dining hall and the souvenir shop are located near the lake to enjoy the beautiful scene. A waiting porch and a canopy on the route of bus going up the mountain are faced to the Taishan Mountain. While the office spaces such as operating offices and VIP acceptance are on the other side of the road. The building strides over the road to form a pass for the convenience of management.

The building form derives from the unique geological characteristics of the Peachblossom Valley, with trenchant edges and corners of a picturesque disorder, just like stones of the Color-stone River. Walking among these "stones", tourists can catch the segment of the great view. Colorful cast-in-place concrete is used to simulate the patterns of stones in the Color-stone River.

摄影：张广源　Photo: Zhang Guangyuan

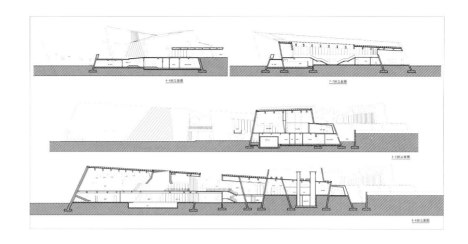
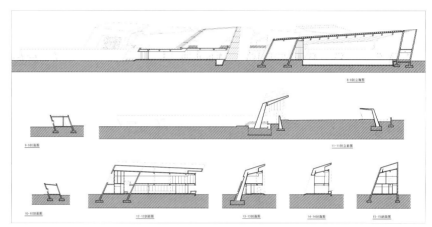

剖面图 / Sections

无锡鸿山遗址博物馆 无锡
Wuxi Hongshan Ruins Museum, Wuxi
2006~2008

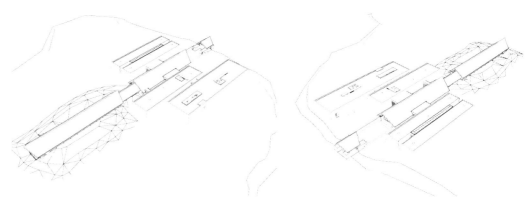

俯瞰图 / Aerial view

鸿山遗址博物馆依托于吴越贵族墓葬群邱承墩而建，用于遗址及文物的研究和展示。

设计重构了一种与当地自然景观相契合，同时反映春秋吴越历史氛围的建筑景象。建筑采用了与周边乡村呼应的地方材料，体量被刻意压低，并将主体部分与封土堆拉开距离。主入口设在西侧，面向主要道路，与遗址景区的游览流线相配合。

建筑造型的来源有三：遗址封土堆的形态；周围呈东西走向的农田肌理；苏南民居的坡屋面。整体建筑形体是一组长方形体量，平行排开并左右错开，草顶土墙，与环境融为一体。只有中部架在门厅和原址上的几段坡屋面被适当突出，融合了朴素的江南民居和粗犷的先秦建筑形态，提示遗址所在的轴线。

建筑外墙为仿土喷砂，局部内凹的空间采用整片玻璃，屋面植草，场地道路采用朴素的石渣铺地等，都力图烘托出具有历史感的古朴悠远的效果；中央公共空间上部的铜瓦坡屋面，以及内墙和部分院落墙体的白色涂料，都概括反映了苏南民居的特点，外墙外侧的土层使建筑与遗址区的封土堆相呼应。

摄影：张广源　Photo: Zhang Guangyuan

The museum as research and exhibit rooms, houses tomb ruins with a history more than 2000 years.

Located in rural fields between riverine towns in southern Chinese, it regenerates an architectural atmosphere, not only getting fit for its natural environment, but also being on a historical scene of the era when tomb was built. Local materials are used to harmonize the museum with surrounding houses, while the building massing is divided into several parts and suppressed to make the tomb ruins standing out.

According to the fabric of background fields, a series of parallel rectangles are placed randomly. Rammed earth walls and greened roofs make most of them being parts of the environments. Only the pitched roofs on the central axis, which are remolded from the roofs of the vernacular residence, are emphasized as leading actors that generate a feature mixing the roughness of ancient architecture and the pureness of local villages.

The primitive simplicity are not only achieved by the simply massing, but also the materials such as blasted sands of exterior walls, scree on pathway, copper tiles and white painting.

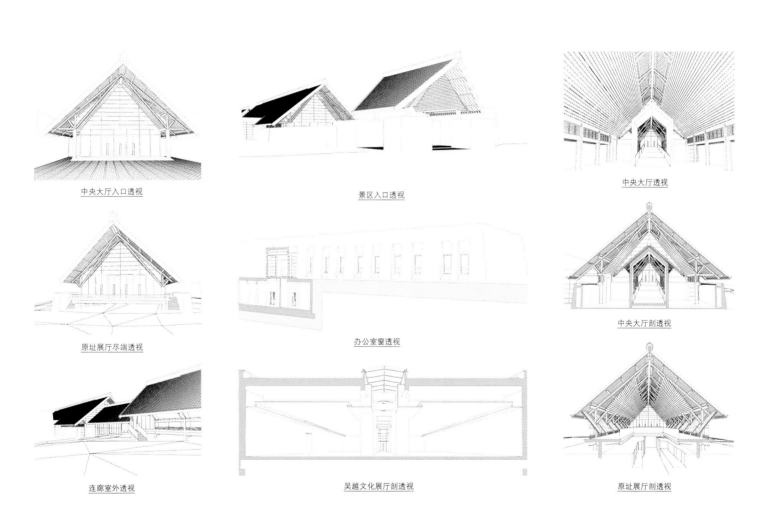

透视图 / Perspectives

前门23号　北京
Legation Quarter, Beijing
2005～2008

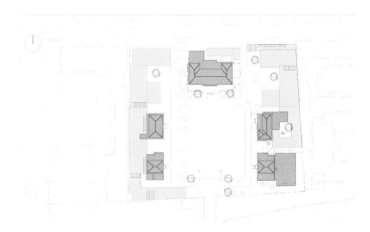

总平面图 / Site plan

项目地处北京市核心区——前门东大街23号，原址为20世纪初美国公使馆所在地，现有五栋北京市文物保护建筑、九棵挂牌古树，多家机构在此办公。改造方案通过对院落的复建和整治，形成包括餐饮、俱乐部、画廊、剧场在内的顶级文化、生活时尚中心。

首先通过恢复领事馆建筑风貌，净化周边环境，突出历史建筑的主体地位。在文保建筑后侧砌筑砖墙，遮挡零碎增建部分，重新恢复一院五楼的原貌，保留古树，拆除低质量旧建筑。恢复重建北大门和北围墙，使之符合东交民巷历史保护区整体风貌的要求。

增加的新建筑采用前低后高、缩小体量的策略，使之处于从属地位。新老建筑内部相连，主入口仍在五个老楼中，保持院落原有格局，给予游人充分体验老建筑历史氛围的机会。新建部分与老建筑相连处采用钢和玻璃的轻型构造，精心保护老建筑的完整性，并以其立面作为新建筑室内空间的主要景观。新建筑在形态上尽可能通透、轻巧，隐在大树之后，克制地表现个体。同时，简约精致的技术和材料也能明显区别建筑的年代，保持历史的清晰度和延续性。

摄影：张广源　Photo：Zhang Guangyuan

Initially built as the American Legation in the beginning of 20th century, NO. 23 of Qianmen East Street housed lots of institutions before the renovation, while its original layout, five independent buildings and a quadrangle, has been damaged after several planless additions. The renovation is focus on restoring the quadrangle and making it a high-grand cultural & fashion center in the center of Beijing city, which provides restaurants, gallery, theatre and entertainment services.

The old buildings are renewed and those low-grade constructions are cleaned up. Brick walls behind buildings screen the disorder parts that incorporate the exterior spaces of the quadrangle with 9 aging trees. According to the preservation regulation of this historic district, the delicate north gate and north wall are restored.

The added volumes step down to make themselves subsidiaries of the old ones. Accesses for them are designed via the five old entries that provide enough chances for visitors to be inspired by the historic atmosphere. Designers juxtaposed this respect for tradition with a distinctly modern vocabulary of steel and glass structural elements, which could not breach the reserved buildings. Hiding behind the shade of trees, the crystal volumes restrain their expression and treat the historical scenes as the primary interior views. The state-of-the-art techniques and materials also give them a distinct feature from the 1920s' style, maintaining the definition of history.

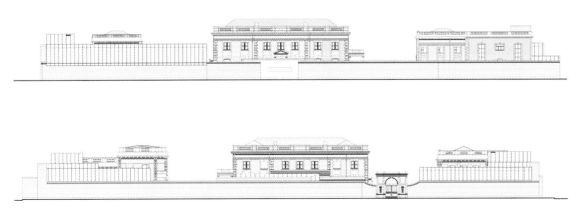

立面图 / Elevations

模型研究 / Model study

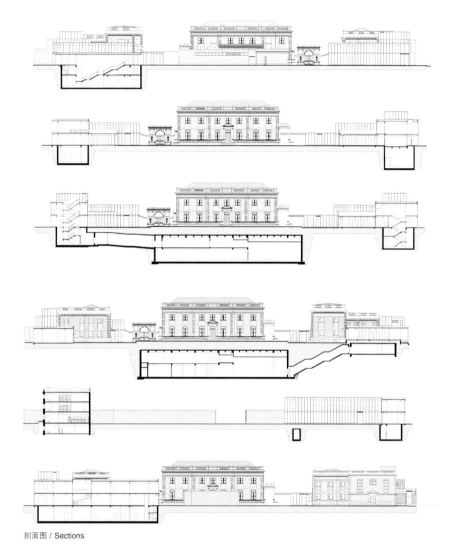

剖面图 / Sections

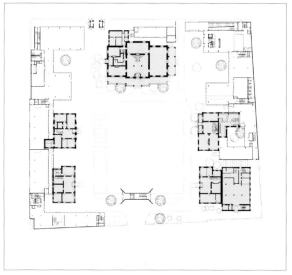

首层平面图 / The 1st floor plan

神华集团办公楼改扩建　北京
Reconstruction of CSEC Office Building, Beijing
2006~2010

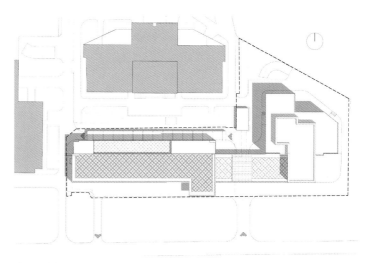

总平面图 / Site plan

项目基地毗邻北二环,在现有建筑西侧有一块因拆除旧建筑产生的空地。其西侧为汉华国际酒店,北侧为住宅小区。在现有的基地条件下,由于受到高度制约,我们采取了将建筑突破原有的基地限制,形体沿二环路伸展,强调建筑长度的策略,这个策略为创造建筑的标志性提供了机会。

考虑到二环路旁的建筑物均强调东西方向轴线,退线整齐,形成了城市隐形界面,我们将办公楼的局部扭转、悬挑,强调南北轴线,突破隐形界面。新建筑悬挑的体量限制在城市绿化带之上,形成一个突出的视觉焦点。新建办公楼结合神华大厦的扩建,将两者连成一体,实现了两者从功能到形象的整合。建筑的中央部分将网络打开,作为新办公楼和原神华大厦的公共入口。外立面旋转45°角框架玻璃网格幕墙为主要维护体系,立面设计上高雅且具有时代特色,区别于一般商业写字楼的风格,并能体现神华公司的企业形象。

CSEC Office Building is located at the north side of northern 2nd ring, very close to the central axis of Beijing. As a reconstruction project, the limitation of site area and building height are the main difficulties for design decision. Instead of demolishing the old building and resetting a new one, the architects suggested that the adjacent high building should also be included into the renewal scheme. Thus, the working spaces can be improved and largely expanded while the various disaccorded buildings are integrated into a unified and meaningful urban interface. The new-added volume develops along the 2nd ring road, twisting and extruding, as the visual focus of the project and connects the new and old organically. The 45 degree angle gridding curtain wall, which acts as the main envelope, endows the building a particular appearance, while an opening in the middle part of the gridding suggests the public entrance.

摄影:张广源　Photo: Zhang Guangyuan

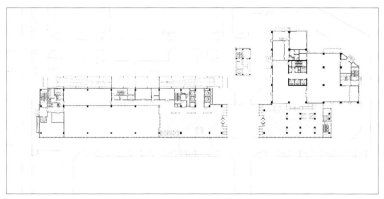

首层平面图 / The 1st floor plan

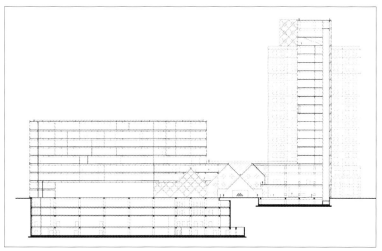

剖面图 / Section

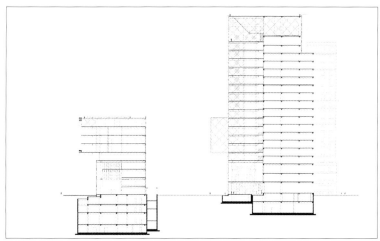

剖面图 / Section

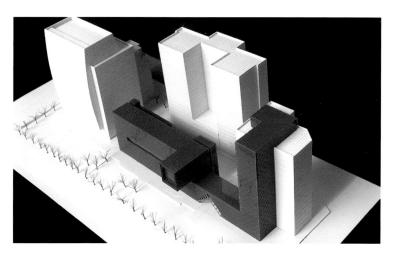

模型研究 / Model study

浙江大学紫金港校区农生组团　　杭州

Agriculture & Ecology Group of Zijingang Campus, Zhejiang University, Hangzhou
2008~2010

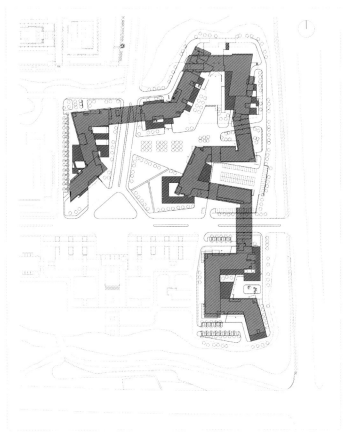

总平面图 / Site plan

中国高校过去十几年的兼并重组始于浙江大学。原杭州农学院并入浙江大学，在紫金港新校区东南隅建设新校舍，功能包括农业与生物技术、环境与资源、动物科学、生物系统工程与食品科学等四个学院和一个国家实验室。教育科研的快速发展使各学院对实验室规模的要求很难确定，而用地的局限又使独立安排各学院建筑会显得拥挤，与整个校区宽敞的绿地环境不协调，于是设计采用了链状布局，将各个学院实验室串接成一个带状灰色板楼，而在一层将每个学院的学术交流和行政管理中心独立设计，呈书院状，与绿化庭院景观相融合。由此达到了科研资源的共享，景观环境的共享，现代性和地域性的共享。

In the past over ten years, there was a wave of merger and acquisition among quite some Chinese universities, and this all started from this university, Zhejiang University. The former Hangzhou Agriculture Institute merged into the Zhejiang University, and new buildings is built at the east-south corner of the Zijingang campus for four institutes and a state-level key lab, namely College of Agronomy and Biotechnology, College of Environment and Resources, College of Animal Science, and College of Biosystem Engineering and Food Science. Due to the rapid development, it is hard to determine the size of each lab; at the same time, due to the space limitation of this campus, it is not suitable for each college to have independent buildings, which won't accord with the landscape. So a chainlike-style layout is adopted, with labs of different colleges linked into a chainlike grey slab-type building, and the academic communication centers and the administrative centers from these colleges are arranged at different floor. The building integrated into the count yard, in which way, the research resources, landscaping resources can be well shared, as well as the modernity and locality.

摄影：张广源　　Photo：Zhang Guangyuan

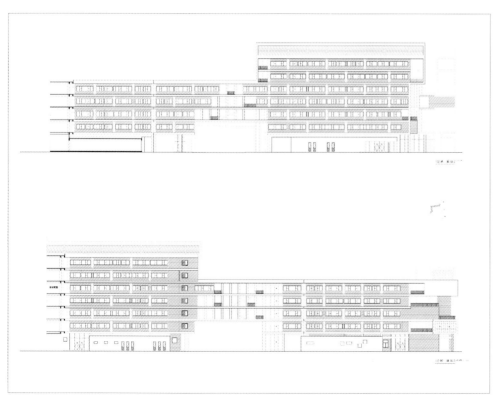

立面图 / Elevations

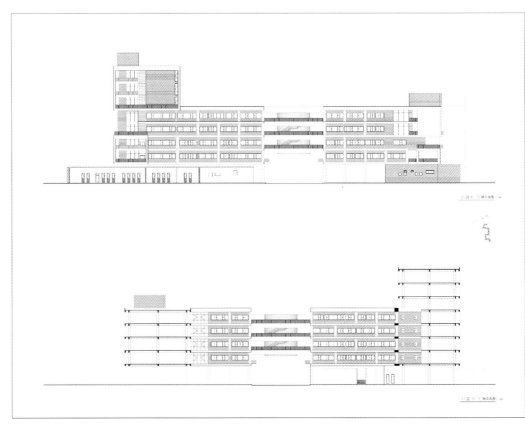

立面图 / Elevations

南京艺术学院设计学院改扩建及南校门广场 南京
Rebuilding of Design College, Nanjing Arts Institute, Nanjing
2007~2009

俯瞰图 / Aerial view

作为对原工程学院主楼的改造和扩建，项目需要营造适合设计专业特点的教学与艺术创作空间，同时利用原有地形特点，解决南校门的交通、停车等功能问题。

三角形的校门应对了城市道路和校园建筑之间的夹角，强烈暗示了朝向北侧山体的空间引导性与复建的原上海美专校门遥相呼应。校门造型简洁，屋顶为透空格梁，坚实的外框内含精巧的玻璃传达室，其幕墙彩色艺术玻璃为校方设计制作。正对校门的大台阶将位于不同标高的校门广场和设计学院广场连成一体，串连整个步行区。利用地形高差增加的可自然通风、采光的半地下车库及小型下沉庭院，活跃了广场气氛。

设计将两栋原主楼、配楼建筑连通，分别容纳教室及展厅。设计拆除主楼西墙的墙裙，植入了空调系统，用铝格栅延续原有的立面外形。主楼东南角加建6层高的教学房间，各房间根据尺度不同可用于展览、教学、会议、制作、休息等用途，成为一座具有提示作用的艺术墙面。建筑东侧出挑一部钢桥，与东侧山体步行道相连。新老建筑之间形成狭长的缝状中庭，空间变化丰富，成为最具活力的交往、活动空间。

摄影：张广源　Photo：Zhang Guangyuan

The genesis of this renovation arose from several aspects. First, the original teaching buildings need to be remodeled as a place which could enhance imaginative atmosphere. Second, the alternation must follow the topographic character of the campus. And the third aspect, resolving the existing functional problems, such as the chaos of traffic and the lack of parking, is also required urgently.

A triangular porch indicates the angle of the city road and the spatial axis for a restored school gate, which has a history ascending back to 80 years ago. The sculptural mass contains an ingenious reception office, which is clad with stained glass window designed by students of the Arts Institute.

Two of the old buildings are connected to house classrooms and exhibition halls. Destroying the west bottom wall, embedding an air condition system, adhering the aluminum lattice to maintain the original form of the building, and adding several vertical stacked classrooms on the southeast corner, the designers change them into a cluster of containers, which can be employed for exhibition, teaching, meeting, manufacturing and resting. Through a steel bridge on the east, students could access the lane on the east hill. Between the old and the new building, a narrow atrium full-filled with various space elements provide opportunities for students and faculty to meet and communicate.

南京艺术学院图书馆　南京
Library of Nanjing Arts Institute, Nanjing
2008~2010

俯瞰图 / Aerial view

南京艺术学院因为是在老校区内进行更新改造，其项目的特点就是见缝插针，左右逢源。图书馆新馆也不例外，需要贴临老图书馆接建，身处现状食堂与教学楼之间的狭长地带，又恰好位于山坡的边缘，地势高差很大，学生由宿舍区到达教学区需要爬高超过10米。

设计将整个建筑上部抬起，下部沉入坡地，中间架空两层，仅有门厅和上部几层与老馆连通，并在尺度上与老馆看齐。大跨度柱廊下的层层台阶和坡道解决了由生活区到公共教学区的步行连通，并在穿行的过程中，不断在视觉和心理上强化这种联系，同时悄悄但持续地诠释了校园的地貌特征。铺满新馆立面的竖向遮阳百叶，不仅使建筑体量显得更为轻巧，使阅览室内的光线更加柔和，更对形成校园的这一处公共空间的完整性、主导建筑群落之间简约利落的整体风格起到了关键作用。

Projects of Nanjing Arts Institute are mainly planned in its old campus, so in most projects, architects have to find ways of solution in a complicated and intensive environment. The project of new library cannot make an exception either. New part of the library needs to be closely connected to the old part, which is located in a narrow area in between a canteen and a teaching building. The site happens to be at the edge of a steep hillside, from which students have to climb over ten meters up to get to teaching area.

The upper part of the library is lifted up, while the bottom part sinks into terraces. Two floors in the middle are elevated overhead. Only the entrance and some floors of the upper part, which are similar in scale, are connected to the old library. Stairs and ramp ways under the colonnade link the living area and the teaching area together. They haven not only emphasized such a linkage in both visual and mental aspects, but also implied the topographic feature of the campus. Vertical sun-shades of the facades made the building look lighter and softened the sunlight in the reading room, and contributed a lot in the formation of a contracted style.

摄影：张广源　Photo: Zhang Guangyuan

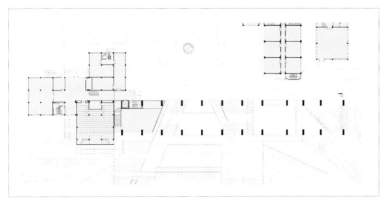
首层平面图 / The 1st floor plan

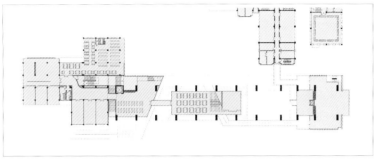
二层平面图 / The 2nd floor plan

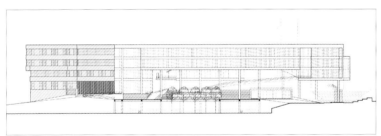
东立面图 / East elevation

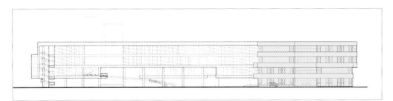
西立面图 / West elevation

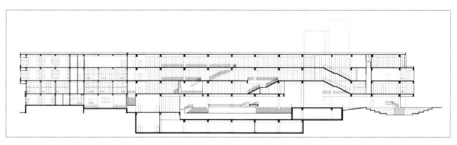

剖面图 / Section

北京外国语大学教学办公楼 北京
Office Building of Beijing Foreign Studies University, Beijing
2007~2009

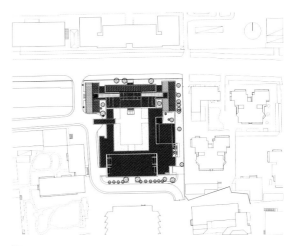

总平面图 / Site plan

用地上原有一栋学生宿舍楼和一栋老图书馆,其中学生宿舍楼是著名建筑师张镈20世纪50年代主持设计的校园建筑的一部分。为了维护校园的整体形态,并与其他保留建筑呼应,设计部分保留了原宿舍建筑的立面,使建筑获得有机生长。

规划布局为内院式,北侧为7层的教学办公主楼,南侧为2层的接待会议部分,东西环廊为校史馆。原有立面及坡屋顶,新建楼体和保留的原有建筑立面及坡屋顶有所区别,在室内外设计中体现新旧对比的戏剧性。设计在原有墙面陈旧的灰砂砖外铺贴灰色面砖,延续其风格。保留"中间走廊、两边房间"的形式,将宿舍的小空间调整为适应办公需求的大空间,在各层设置会议室,并增设两部楼梯、两部电梯。新建楼体向南侧扩建,通过落地玻璃窗获得更多的日照。

配楼强调与主楼的关联性,一层与连廊融合,以一道长长的灰墙与主楼相连,二层则以混凝土体块的形式与灰色墙体脱开,塑造纯朴的意境。内庭院的设计提炼了传统合院建筑的意境,使用较多的空地,辅以北京常见植物和传统圆铜缸,创造出优雅娴静的环境。

摄影:张广源 Photo: Zhang Guangyuan

The reconstruction strategy focused on restoring a sense of wholeness to the campus's original master plan, designed in 1950s by Zhang Bo, a famous architect of China. As a part of architectural heritage, the traditional C-shaped dormitory is maintained and regarded as a witness of the organic regeneration course.

A 7-story slab is added in front of the original building, facing a new quadrangle to the south. The 2-story building, which defines the south edge of the quadrangle, contains reception and conference functions, while the history exhibition has its home in the east and west galleries.

Dramatic contrast is imposed between the original and additional structures. Attached with grey ceramic tiles, the old brick walls continue their classic style with several recovered pitched roofs. Although the small dorms are enlarged to meet requirements of modern office environment, the double-loaded corridor layout is maintained. In the same time, every level is equipped with public meeting room, as well as two elevators and two stairs which are inserted as new vertical traffic cores to anchor the old building. The additional expressed structural framework, which follows the proportion of those old structures in the campus, and features floor-to-ceiling windows to the south to ensure enough sunlight.

A flat box, which inherits the palette of the main building, is raised by the three-side enclosure of the quadrangle. Projecting outward, it makes itself floating above the grey wall, with the contract between pure concrete feature and delicate texture of grey bricks. Rooted in the indigenous culture of old Beijing, the quadrangle is not furnished with too much landscape components. Normal plants and brass vats, which are restrainedly placed, imitate the peaceful scene of traditional courtyard.

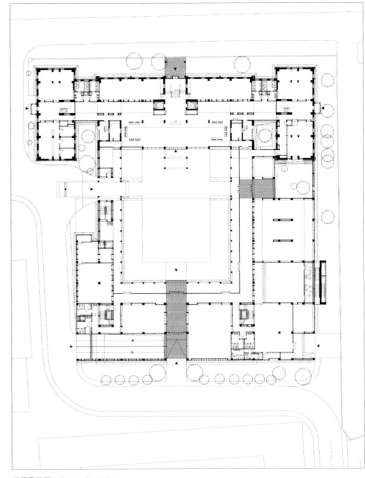

首层平面图 / The 1st floor plan

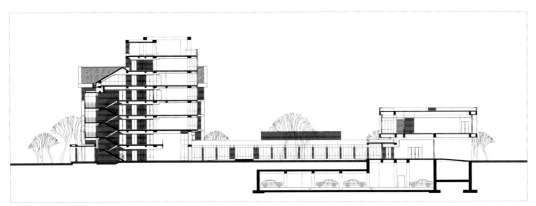

剖面图 / Section

北京外国语大学综合体育馆　北京
Beijing Foreign Studies University Complex Gymnasium, Beijing
2005~2008

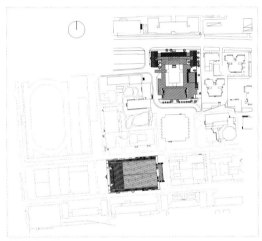
总平面图 / Site plan

紧邻体育场的用地狭长而紧凑，使得各项设施不能处于同层。设计通过垂直空间组合，使得各项功能得到适宜布置，尽可能缩小了占地面积，留出更多的室外空间。

综合体育馆、游泳馆入口布置在一层。中途训练馆布置在综合体育馆的西侧，与综合体育馆使用上可分可合，方便教学、比赛使用；健身房和艺术沙龙布置在游泳馆上空南北两侧。体育教研室放在可俯瞰体育场西侧，便于管理。

南北两侧2.7米宽的混凝土柱廊弱化了建筑边界，降低了建筑的压迫感。设于其中的室外楼梯除满足疏散要求外，也增加了建筑内外的交流，成为立体的校园路。结合东侧绿化广场设计的挑台，增强了体育馆与周边环境的互动联系。西立面则通过设置遮阳板降低太阳西晒的影响。建筑的立面色彩和语言延续了临近的逸夫教学楼的形态，形成了积极的对话关系。

A proximate site to the sports field arouses the contradiction of limited space and multiple functions. Vertical stacked space composition is used to accommodate every function adequately.

The comprehensive gymnasium and the natatorium are on the first floor. The training gym lies to the east of the comprehensive gym; it can merge into or separate from the comprehensive gym for flexible uses. Above the swimming pool, there are fitting room and artistic club seated on the south and north. Overlooking the sport filed is a special advantage for the teacher's office on the west.

Two grand arcades with 4-story height and 2.7m width obscure the straight edge of the huge mass. Besides satisfying the evacuation regulation, the outdoor stairs are considered to be three-dimensional paths for communication. The huge balcony on east facade, facing a landscaped square, extends the interaction between the gym and its surrounding. Vertical shafts cantilevered from the west facade, serves as sunshades to protect the offices from harsh sunlight. Its design continues the themes of the existing building – Yifu Building, which is not far away, combining the simple forms with the same vocabulary and color.

摄影：张广源　Photo：Zhang Guangyuan

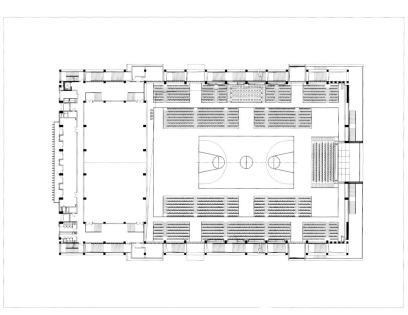

四层平面图 / The 4th floor plan

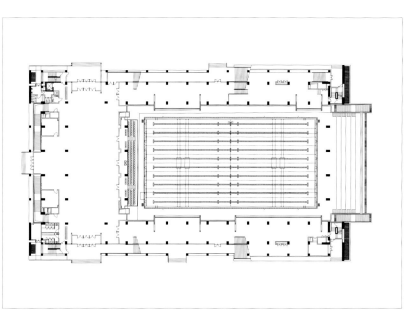

首层平面图 / The 1st floor plan

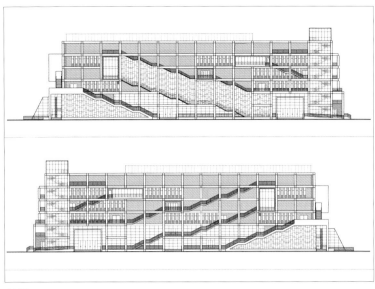

立面图 / Elevations

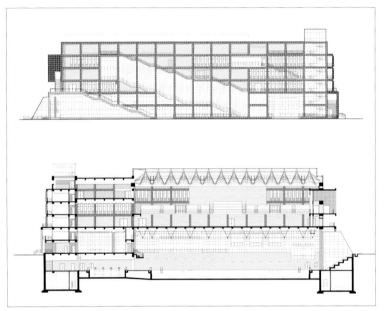

剖面图 / Sections

承德城市规划展览馆 承德

Chengde City Planning Exhibition Hall, Chengde
2008～2009

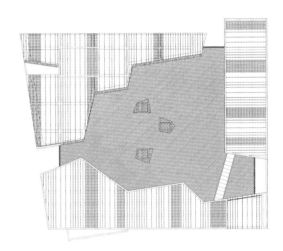

屋顶平面图 / Roof plan

作为重要的城市公共文化设施，承德城市规划展览馆的设计既要体现承德的历史文化和地域性特色，又需要展现城市建设的成就和美好未来。

设计方案从展览馆最重要的城市规划模型出发，将建筑体量切割成三部分，隐喻承德独特的地形和城市脉络。建筑沿街面与展览馆中的城市模型布置方向呼应，从山地环境抽象而来的类似岩石的材质和体形，从避暑山庄、外八庙等古建筑中提取的红墙和渐变方窗等传统建筑元素，覆盖于主展厅的外立面和屋顶。建筑内部空间亦以城市规划模型为中心，使参观流线和办公空间有机结合在其周边。相较周边建筑，展览馆没有过长的沿街面和突出的高度，却通过突出建筑的整体感和与城市空间、城市地貌及文化特色的结合，充分表达了与城市的和谐关系。

建筑体形方正，入口位于悬挑一角下的大台阶处。立面通过成组悬挂深色石材条板，以不同宽度的间距突出建筑的体量感；展览部分外墙和屋顶为红色，开有下疏上密的方洞口，与普陀宗乘之庙的外墙颜色呼应。三块深灰色的建筑实体包住红色展览部分，两者的结合表达了承德传统建筑隐藏于群山中的意境。

摄影：张广源   Photo: Zhang Guangyuan

An important public culture facility, the City Planning Exhibition Hall is designed to reflect the local culture of the ancient Chengde city, and presents the achievements of city construction today.

The integral volume is divided by three cracks originated from the city planning model in the center of the hall, suggesting the unique topography and the urban texture of the city. The mass enveloped with grey intense lattices, which borrow the window patterns of ancient temple, and allude to the surrounding mountains in both texture and form. The intricate play of light and shadow changes with the sun, particularly emphasizing the dark red inner core, a container of the most important part of the hall – the city planning model. The contrast between grey and red, textured and smooth, also indicates the partnership of traditional buildings and their environments. To the interior, visit circulation and office rooms are arranged around the model.

Compared with the neighboring buildings, the exhibition hall doesn't have a great mass, but, with its corporate image and its understanding of local characteristics, the building really becomes a focal point of the city.

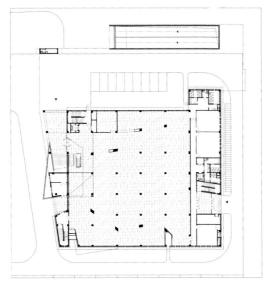

首层平面图 / The 1st floor plan

二层平面图 / The 2nd floor plan

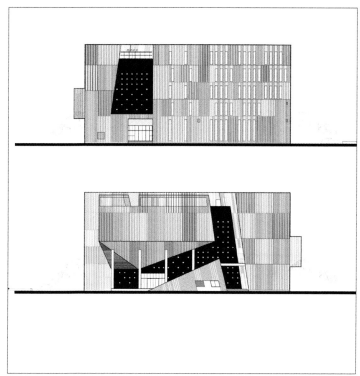

立面图 / Elevations

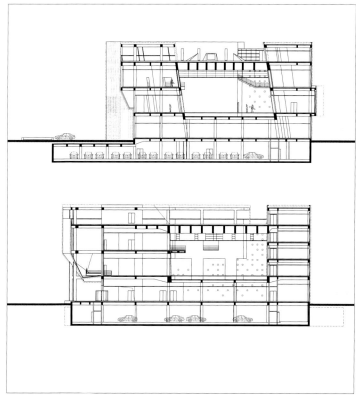

剖面图 / Sections

崔愷访谈

采访人／黄元炤
北京 2011.9.27

黄：就我的观察，您其实在本土设计基础上，用一个持中态度接受不同设计语言的思考与态度，仍然是在功能性考量下，体现作品多元多面貌的形态，所以，您的"本土设计"范围是非常广义的，其实是总结所有设计的思考。

崔：很广义的。当我起了这个名字以后，重新去解读我自己的设计，实际上都是跟某个特定场所、跟特定环境的地域文化有关。"本土设计"就是这方面。

黄：我所理解的本土化是在主体不明确、且思想不清楚的情况下，对应于全球化在世界上日渐趋同产生的一个反义词，或者说是对抗的手段，有种临时性与被迫性，也是另外一种倾向于文化（本土与外来）表述的语言、趋势或潮流，而"本土设计"是您个人对外表述的设计思想与说法，想必您也想要把它作某方面的提升？

崔：实际上起初我说"本土设计"时，一些外国朋友不太能理解。觉得你是一个有国际视野，或者有开放心态的建筑师，为什么会谈"本土设计"？有一个法国记者，听我讲"本土设计"时，我说我的"本土设计"不是排外、不是市场封闭，她说是吗？你真确定你有这样一个想法？可能是因为当时我用的是"Native design"这种译名，让人觉得是一个跟种族、领土有关系的说法。用"Native"这个词，我自己也觉得有点不准确，后来用"本土设计"当主题参加过几次国际论坛演讲，用这个词的时候我都要跟别人解释，觉得有点别扭。有次朱剑飞老师来拜访我，我说本土设计这个词的英文我一直很纠结，我想一定要在国际交流当中让人明白你说的是什么，不要有误会。他回去后给我发邮件，改了几个译词，最后我选了一个词叫"land-based rationalism"，反译回来是"以土地为本的理性主义"。讲到理性主义，在设计当中重视理性思考是现代建筑发展中很重要的一个基本思想。

黄："理性主义"都带有一点古典性格，一开始我们所说的理性主义可能是密斯之类一路演变过来的，或者意大利新理性主义。

崔：我觉得是一种思维方式，讲道理、有逻辑。在密斯时代，理性主义是关注到工业化生产方面，更多的是工业化、标准化与极简主义，从这方面发展出一种新的美学。今天的理性主义更多是关心环境、人文这些方面，可能跟那时候在具体倾向上不太一样。"land-based"，立足于土地的理性思维，"land"在土地与地域方面的含义比较容易理解。

黄：传统、文化与地域都包括在这里面了。对其他人来说，其实有一点点矛盾，我们通常说"本土设计"，观念上都会觉得是属于某个地区范围内，关注当地环境、历史、文化等方面的思考与设计，听您说来，您这个"本土设计"在中国境内的设计思考下，解释是相对来说的广义，这有一种您想向别人解释清楚对这一概念彼此认知差异的意思，所以，您解释起来会特别累，您认为您做的非常广义，可是人家会认为就是某个地区范围内，偏向于狭义解释的"本土设计"。

崔：我觉得有些事总想更容易让人理解。我可以把这四个字拆开来解释："本"肯定是本质、本源、根本，跟理性主义的基本立场是一致的；"土"，当然是强调地域、强调资源、强调环境；把"设计"两字也拆开了，"设"，实际上就是设置一个策略；"计"，实际上就是策略、计谋，用这四个字的中文解读方法把它变成一个比较通俗的道理。

黄：我在您所提出的解释中发现，不仅包括设计，也包括计划这部分。就是说，您似乎认为建筑设计是一项综合性服务事业，不单单只是主观地表达个人情感的纯设计，似乎还有客观所需考量的成分在，必须综合考量其他环节。

崔：对。我有一个观点，今天在大规模建设当中，最重要的是原则和立场以及使用的策略，这个正确与否是很重要的，也是最基本的。在这个之下发展出来的造型、表皮、建构，都是在这个基础上的更高的艺术层次。对我来讲要应付比较多的设计项目，我觉得最起码要做到策略是对的。

黄：就是跟您所处的环境有关系。

崔：比方说这个建筑需要向城市开放，那这个立场在整个设计当中是贯穿始终的，至于具体的设计方法，我觉得是可以有弹性的。有时候策略对了，只是完成度不够，但比较踏实的是这件事做对了，只不过建筑不那么完美、不那么艺术。反过来说，有些建筑师把自己的建筑做成艺术品，确实做得非常好，但是往往跟城市的关系是不对的，对应该关心的基本问题反而忽视了。造成长久的难以弥补的遗憾。所以我觉得应该提倡重新回到理性思考的路径上来，建立起建筑师基本的价值观和责任感。

黄：未来您如果被写入建筑史，肯定还是会归属于一个基本的流向或者流派，历史学家会定义你的。而您所提出的"本土设计"可能就会被表述成植根于地域文化中的设计思考，您的作品中有部分也具有地域性的特质，您曾思考过运用当地的民族文化资源作为建筑设计语言的一种转换与呈现吗？

崔：也有人给我提过问题，问：你这个跟"地域主义"到底什么关系？"地域主义"在我看来是在现代主义建筑发展的鼎盛的时期,在国际式风格到处蔓延的时候出现的。

黄：在20世纪四五十年代发展与蔓延开来。

崔：总的来说它多呈现在一些欠发达地区的建筑活动中,如印度、墨西哥等第三世界国家,是主流之外所发生的现象,所以理论家赋予它一个名字——"地域主义"。但我不觉得中国应该被边缘化,我们一直试图重新回到主流,但实际上这也须有个过程。

黄：中国被世界边缘化已经两三百年了。

崔：但是作为中国建筑师来讲,我不希望我们提一个口号是自我边缘化的。我觉得从概念的基本出发点来讲,我们中国建筑迟早会站在中间的位置上,而不是总把自己边缘化,就是这样的一个想法。另外也有人问"本土设计"跟"文脉主义"有什么关系？我认为"文脉主义"比较关注对现代城市的反思,重新研究城市尺度的问题和与传统城市的空间价值。但后来它转向一些视觉与形式符号,过于形式化,最后走进了死胡同,不再有创造力,也很快过时了。我觉得"本土设计"不应该单纯强调形式,其关注的也远远不止形式问题。如果说"文脉主义"跟我们在学术思潮上类比的话,那就是20世纪七八十年代争论不休的民族形式的问题：形似呀还是神似？

黄：可能也不能只说神似,因项目当中还有形式。

崔：形式实际上不是关注的重点,而是产生的结果。有些时候可能需要具象一点,有些时候抽象一点,也有时候完全不需要从形式上去解读环境。

黄：您虽然强调本土设计,但若先撇开本土设计的论述,就我对您的观察,您的设计路线似乎一直在变,在满足功能性思考的前提下,近期作品变得更趋近于干净化、简单化的倾向,如山东广电中心、北京西山艺术工坊,但体量关系还是很明显,而这改变似乎还一直处在摆荡与摸索的状态,无法从单一的作品表述中看出是您的设计。

我换个问法,建筑实然是一个感官物体,它也代表着建筑师个人的标识性语言与符号。就传统来说,20世纪以前,吸引人的是建筑的形,是一个广泛的约定成俗的形式,且不管东方与西方,在宫殿、庙宇、亭、台、楼、阁,似乎较少从建筑的形当中去关注建筑师个人建筑语言的表述。在现代来说,20世纪二三十年代一直到六七十年代,建筑的"形"与人似乎做到某种程度的结合,如柯布、密斯、赖特、阿尔托、高迪等,到晚期现代主义中的尼迈耶、康、保罗·鲁道夫等。再到了后现代时期,由于建筑又回归到寻找传统文脉、注重象征与符号的语言,似乎建筑师的个人语言表述又模糊了。到了解构主义刚开始成形时,建筑的"形"与人似乎又做到某种程度的结合,如哈迪德、盖里、埃森曼。到解构主义后期,建筑的形与人又模糊了。

所以,就现代建筑发展史来看,初衷实然是建筑物就代表其设计人,这是现代建筑的一个开始,也是与传统不同的地方。而就我的观察,您的设计时期正好处于世界建筑史发展中的后现代与解构主义时期,所以,您的作品一直有着这两种时期的特征,一下子与真实清晰的结合,一下子又模糊了,您似乎在真实跟模糊之间游走着,您自己怎么看待这个现象？您是否仍在寻求设计上的自明性？

崔：这是我特别要强调的一个设计原则,就是我不认为建筑是属于个人的作品,这跟其他艺术很不一样。

黄：因为以往艺术作品被认可,作者往往已死,艺术家总是在作品生产到创造过程中逐渐自我消亡,然后作品就独立于一切之外,只剩下作品的形式。

崔：刚才您从建筑史角度对建筑师的归纳确实很有意思,有些建筑师强调建筑跟自己的关系,可以叫"图章式建筑师"。不管在哪儿做,风格是完全一致的。就像你说的,可能这个社会发展到一个阶段,把建筑当成是一种表现,可能跟艺术有关,跟消费文化有关,跟品牌有关。以前教堂就应该是这样的,不管谁设计都是这样,只不过谁设计得更好。而现在流行消费文化,品牌变成了很重要的载体。

我自己对这个有不同的看法。首先我觉得要产生一个特别有价值的、或者特别有独特性的视觉语言是非常不容易的,并不是所有人都能做到。因为他们创造了一个建筑学的艺术语言,带动了一个时期建筑的发展,让所有人追随他,引导了潮流,这是极少数建筑师才能达到的高度。不能要求所有建筑师都能这么做。

另外还是回到"本土设计"的问题,我觉得建筑产生于特定场所,建筑跟谁来做这个设计,按理说没什么关系。当然每个人对这个场所的解读会不一样。我一直希望是一个客观的角度,我每次判断这件事的时候,我给大家讲道理的时候,都讲的是实实在在客观的道理,而不是我觉得怎么样。有些建筑师跟我讨论时说我觉得如何如何,我还纠正他,我说,你陈述你在现场看到了什么,不要太主观。我个人认为"本土设计"的基本立场,来自于对场所真正的发自内心的尊重,所以,建筑呈现的形态应该更多的是去表达这个场所本身应该有的一种状态,我觉得不应该片面强调个性化,尤其是不应该为了个性去做一些不恰当的表达,这是我的看法。

Interview

Beijing 27th Sep.2011

Huang: According to my observation, you are quite neutral in accepting different thoughts and attitudes in design base on the land-based rationalism. In fact, you are still trying to express the variety of works on the premise that function requirements are fully considered, which means the land-based speaking you mentioned is a quite general idea,and it is essentially the conclusion to all design.

Cui: Yes, it's a general idea. After I named this idea, I reexamined my works, and found out that they all connected with certain location or certain local culture. That is Land-based Rationalism.

Huang: As far as I can see, native is the antonym to globalization and a method against it and there is no clear subject or thoughts. It's temporary and compelled and is a language or trend of a culture (native or foreign). Land-based Rationalism is how you express your design thought and philosophy, are you planning to raise it somehow?

Cui: In fact, at the beginning when I introduced this idea of native design, some foreign friends didn't quite get it. They wondered why an architect with international view and open attitude like me would talk about this. There was a French journalist with such opinion, and I explained that I am not advoacacy of anti-foreign or market closure. She still feels suspicious. I think maybe it is because I chose "Native Design" as the translation for my idea, which may mislead people to nation and territory related issues. I myself feel not quite good with the word "native", especially when I have to explain myself every time I took part in international forums and used it as my topic. One day, Mr. Zhu Jianfei visited me, and I told him my confusion on this word, and my hope that a clear understanding could be achieved during the international communication. He then mailed me and provided me some words for replacement. I then chose "land-based rationalism", which means a rationalist based on land. Rationalism has been an important thought in the development of modern architecture.

Huang: In nature, rationalism is classical more or less. The rationalism may start from Mies van der Rohe or the neo-rationalism in Italy.

Cui: I think rationalism is a mode of thinking,or to think in logic. In the period of Mies, rationalism was more about the commercial process. It concerned more about commercialization, standardization, minimalist,and so on. A new esthetics has developed over this. The rationalism today is more about environment and humanity. It is different from what it was.

Huang: It is also about tradition, culture and region.

Huang: For people that not engaged in this field, it may be a little inconsistent. Usually, when we talk about land-based rationalism, we always think it is about thoughts and designs that focus on local environment, history and culture, and it is relatively a narrow idea. But your land-based idea is a relatively general idea of thoughts on design in China. There is a difference between what you try to explain to others and what information others get after your explanation. So there will be quite some

work for you, you think you are talking the more general one, while others will think you are talking about the narrow idea instead.

Cui: I think there is a way to make it more understandable. I can explain this idea with four words: nature, local, strategy and stratagem. Nature, emphasizing it is accord with traditional rationalism; local, emphasizing the region, resources and environment. Strategy and Stratagem are the understanding of design in Chinese word meaning.

Huang: In your explanation, I found strategy and stratagem as well as the design itself, it seems that you think architecture is a comprehensive service business, it is not only the pure design to reflect designer's feeling, but also the consideration of other fields.

Cui: Yes, I do held a viewpoint that under the circumstance of massive construction, it is important to have a right principle and strategy. The modeling, appearance or structure are just some art based on this principle and strategy. As for me, since there are many projects in my studio, I think at least I should ensure a right strategy.

Huang: You are talking the environment you are in.

Cui: For example, architecture should open to the city. This is the basic position that need to be kept during the whole design process. As for the exact design methods, I don't think there in an exclusive answer. Sometimes, there is a right strategy that is not fully achieved, but the fundamental aim is achieved, that is good. Only the building is not so perfect, not so artistic. On the contrary, some architects, they always make their works some artwork, which are really quite artistic, but the relationship between the buildings and the cities are not right, because the fundamental problems are neglected, which will arouse a real pity that can not be made up. So I think we should get back to the rationalism, and rebuild the fundamental sense of worth and responsibility of architects.

Huang: In the future, if there is a chapter about you in the architectural history, you will be labeled to a certain school. Maybe the historians will define your idea of land-based rationalism as a thought based on regional culture. In some of your works, there is a regional characteristic, and you did take some regional culture resources into your design language.

Cui: There did some people ask me this question, what is the difference between your land-based rationalism and the regionalism. Well, for me, I think the regionalism arouse in the golden age of modernism when international style has been introduced everywhere.

Huang: That is 1940s and 1950s.

Cui: In general, the regionalism was popular in some under-developed countries, such as India and Mexica, which was outside of the mainstream. That is why it is called regionalism. But I think China should not be peripherized, We are trying to get back to the main stream, and that need some time.

Huang: China has been peripherized for over two or three hundred of years.

Cui: As a Chinese architect, I don't hope that we raise a self-peripherized thought. I think sooner or later, architecture in China will be back to the mainstream. We can not always be peripherized. There were also some other people ask me what is the relationship between the land-based rationalism and contextualism. I think contexturalism is more about the reflexion to modern cities, which tries to restudy the scale of cities and the traditional urban space value. But later it became too formalized, and emphasized too much on visual and formal symbols.That is why it lost its creativity and became out-fashioned. I think land-based rationalism shouldn't pay too much attention on shapes and should care more than just shapes. If we want to give an analogy to the difference between contexturalism and land-based rationalism, the disputation of nationalistic form in 1970s and 1980s could be one, and the difference is whether it should be similar in shape or in spirit.

Huang: Maybe not only similar in spirit, since there always a shape in projects.

Cui: Shape is not the main concern, but the result. Sometimes it should be more concrete, sometimes it should be more nonfigurative, and there are also times no need through shape to understand the environment.

Huang: Even though you have been talking about land-based rationalism, but if we let alone the definition of land-based rationalism, according to my observation, your design course has been always changing. Based on satisfaction to the requirements of function, your later works has becoming more clean and simple, such as the Shandong Broadcasting and Television Center and the Plot B of the Inside-out. But the shape relationship is still quite obvious. And this changing has been unsteady. One can not tell your design characteristics from a single building.

Let me ask my question in another way. Architecture in fact is sensorial object. It is also a significant language and symbol of architect himself. Before the 20th century, the shape of buildings are more in a conventional shape, both in the eastern or the western society, such as palaces, temples, pavilions, terraces and open halls. The personal architecture language of architect himself is seldom expressed. From 1920s or 1930s to 1960s and 1970s, the shape and the personal architectural characteristics are somehow combined, such as the design of Le Corbusier, Mies, Wright, Aalto, Gaudi, and so on. And later there is Niemeyer, Kahn, Paul Rudolph, the White and the postmodernism. In the postmodernism period, because architecture came back to traditional context, and pay attention to symbol and signification, the personal characteristics became vague. When deconstruction emerged, the shape of architect and the personal characteristics are combined in some way again, such as Hadid, Gehry,

Eisenman. But lately, when the deconstruction developed, it became vague again.

So according to the modern architecture history, architecture represents the architect himself.This is how the modern architecture started, and this is the difference between modern and traditional architecture. When I overviewed your works, I found that these works are during the period between the postmodernism and structuralism, and these works do carry characteristics of these two different periods. Sometimes the combination is very clear and true, but sometimes the combination is very vague. It seems that you are wandering between these two. How do you judge this phenomenon? Are you still trying to figure out your self-evidence?

Cui: This is a design principle I have always being emphasizing. I don't think architecture is personal work. Architecture is quite different from other categories of art.

Huang: This is because when a work of art is approved by public, the author often passed away already. The artist himself gradually dies out in the procession of his work, and then the work become independent and only the shape of the work exists.

Cui: Your conclusion of architects according to the architecture history is quite interesting. Some architects intensify their relationship with their works.I call them stamp-architect. No matter where the projects are located, they shares the same pattern and style. Just like what you said, maybe in a certain period of society development, architecture is considered as an expression,and maybe it is related to art, to consumption culture, or to brand. Take churches as an example, in the old days, no matter who design a church, it should be alike. The only difference is who makes it better. While in nowadays, due to the popularity of consumption culture, brand becomes a very important carrier.

I personally think differently. First of all, it is not easy to have a valuable and unique visual language, which is not achievable for anyone. Because to create a language of architecture, which push forward the architecture, and everyone follow, this is what only few architects can do.You can not ask every architect to work like this.

Let's back to land-based rationalism, I think architecture grows from the certain place. There should be nothing do to with who is the designer. Of course understanding to the place can be different for different people, but I hope it is from an objective angle. Everytime I make judgment or when I tell princiles to others, I alway talk about the objective reasons instead of what "I thought". Sometimes when an architect told me "I think", I will correct him to just tell what he has seen on that site and don't be subjective. Personally, I think the fundamental position of land-based rationalism is the respect to the site from one's heart. So the shape of an architect should express more about this place itself. I don't think individuation should be over emphasized, or even improper expression, and that is my opinion.

作品年表
Chronology of Works

北京外语教学与研究出版社办公楼

工程地点：北京市海淀区
结构形式：钢筋混凝土框架结构
建筑面积：1.60万平方米
竣工时间：1998年

合作建筑师：郑莹、林红

外研社二期工程(印刷厂改造)

工程地点：北京市海淀区
结构形式：砖混、框架结构
建筑面积：0.35万平方米
竣工时间：1999年

合作建筑师：王祎

现代城高层公寓

工程地点：北京朝阳区
结构形式：钢筋混凝土剪力墙结构
建筑面积：20万平方米
竣工时间：2000年

合作建筑师：吴霄红

北京外国语大学逸夫教学楼

工程地点：北京市海淀区
结构形式：钢筋混凝土框架
建筑面积：1.13万平方米
竣工时间：2001年

合作建筑师：王祎、商伟玲

中国城市规划设计研究院办公楼

工程地点：北京海淀区
结构形式：钢筋混凝土框架 局部无梁楼盖
建筑面积：1.41万平方米
竣工时间：2002年

合作建筑师：李凌

水关长城三号别墅

工程地点：北京延庆县
结构形式：钢结构和钢筋混凝土结构
建筑面积：0.039万平方米
竣工时间：2002年

合作建筑师：徐丰、刘爱华

清华科技创新中心

工程地点：北京市海淀区
结构形式：框架-剪力墙结构
用地面积：1.60万平方米
建筑面积：5.90万平方米
设计时间：2001年
竣工时间：2002年

合作建筑师：林红、柴培根

富凯大厦

工程地点：北京市西城区
结构形式：核心筒框架结构
用地面积：1.10万平方米
建筑面积：11.90万平方米
设计时间：2001年
竣工时间：2002年

合作建筑师：崔海东、张波

北京雅昌彩印天竺厂房综合楼

工程地点：北京市顺义区
结构形式：框架结构
用地面积：1万平方米
建筑面积：1.68万平方米
设计时间：2001年
竣工时间：2004年

合作建筑师：汤钧、许路

外研社大兴国际会议中心

工程地点：北京大兴区
结构形式：框架结构
用地面积：8.90万平方米
建筑面积：8.90万平方米
设计时间：2003年
竣工时间：2004年

合作建筑师：王祎、商玮玲

广东东莞松山湖商务办公小区

工程地点：广东省东莞市
结构形式：框架结构
建筑面积：12万平方米
设计时间：2002年
竣工时间：2005年

合作建筑师：任祖华、丁力群、杨金鹏

西安紫薇山庄度假村及别墅区

工程地点：陕西省西安市
结构形式：框架结构
用地面积：70万平方米
建筑面积：15.5万平方米
设计时间：2003年
竣工时间：2004年

合作建筑师：李佳玲、蒋成钢

北京蓝堡国际公寓

工程地点：北京市朝阳区
结构形式：框架剪力墙结构、部分框支结构
用地面积：1.5万平方米
建筑面积：13.4万平方米
设计时间：2002年
竣工时间：2004年

合作建筑师：吴霄红、张萍萍

民航总局办公楼改造

工程地点：北京市东城区
结构形式：框架剪力墙结构
建筑面积：2.4万平方米
设计时间：2001年
竣工时间：2004年

合作建筑师：王祎、商玮玲

大连软件园九号办公楼

工程地点：辽宁省大连市
用地面积：1.50万平方米
建筑面积：4.20万平方米
设计时间：2003年
竣工时间：2004年

合作建筑师：李维纳、丁峰

首都博物馆

工程地点：北京市西城区
结构形式：下部钢筋混凝土框架-剪力墙
　　　　　上部钢屋盖
用地面积：2.40万平方米
建筑面积：6.30万平方米
设计时间：2002年
竣工时间：2005年

合作建筑师：崔海东、汤钧

殷墟博物馆

工程地点：河南省安阳市
结构形式：框架结构
用地面积：0.65万平方米
建筑面积：0.35万平方米
设计时间：2005年
竣工时间：2006年

合作建筑师：张男

北京德胜尚城

工程地点：北京市西城区
结构形式：框架结构
用地面积：2.20万平方米
建筑面积：7.20万平方米
设计时间：2002年
竣工时间：2005年

合作建筑师：逄国伟、刘爱华

辽宁五女山遗址博物馆

工程地点：辽宁省桓仁县
结构形式：框架结构
用地面积：15.90万平方米
建筑面积：0.30万平方米
设计时间：2003年
竣工时间：2008年

合作建筑师：张男

北京市数字信息出版中心

工程地点：北京市东城区
结构形式：框架剪力墙结构
用地面积：1.04万平方米
建筑面积：4.90万平方米
设计时间：2004年
竣工时间：2008年

合作建筑师：何咏梅、林琢

山东理工大学图书馆

工程地点：山东省淄博市
结构形式：框架结构
用地面积：1.81万平方米
建筑面积：3.67万平方米
设计时间：2005年
竣工时间：2007年

合作建筑师：柴培根

拉萨火车站

工程地点：拉萨市
结构形式：钢筋混凝土框架结构，局部钢结构
用地面积：11.2万平方米
建筑面积：2.40万平方米
设计时间：2004年
竣工时间：2006年

合作建筑师：单立欣

中国驻南非大使馆

工程地点：南非开普敦
结构形式：混凝土框架结构
用地面积：2.47万平方米
建筑面积：1.16万平方米
设计时间：2005年
竣工时间：2012年

合作建筑师：单立欣、喻弢

大连软件园软件工程师公寓

工程地点：辽宁省大连市
结构形式：框架剪力墙结构
用地面积：1.40万平方米
建筑面积：2.80万平方米
设计时间：2005年
竣工时间：2006年

合作建筑师：曾筠

韩美林艺术馆

工程地点：北京市通州区
用地面积：1万平方米
建筑面积：0.90万平方米
设计时间：2004年
竣工时间：2008年

合作建筑师：吴斌

西昌文化艺术中心及火把广场

工程地点：四川省西昌市
结构形式：框架剪力墙结构
用地面积：13.83万平方米
建筑面积：2.54万平方米
设计时间：2005年
竣工时间：2007年

合作建筑师：何咏梅、李斌

奥运多功能演播塔

工程地点：北京市奥林匹克公园
结构形式：钢结构
用地面积：0.63万平方米
建筑面积：0.43万平方米
设计时间：2007年
竣工时间：2008年

合作建筑师：秦莹、张军英、傅晓铭、康凯

奥林匹克公园下沉庭院3号院

工程地点：北京市奥林匹克公园
设计时间：2007年
竣工时间：2008年

合作建筑师：傅晓铭、石磊
景观设计师：史丽秀、巩磊

天津博学院

工程地点：天津市静海县
结构形式：框架剪力墙结构
用地面积：1.30万平方米
建筑面积：1.23万平方米
设计时间：2006年
竣工时间：2009年

合作建筑师：单立欣、何理建

苏州火车站

工程地点：江苏省苏州市
结构形式：钢筋混凝土框架结构和密排菱形空间桁架结构
建筑面积：8.57万平方米
设计时间：2008年
竣工时间：2011年

合作建筑师：王群、李维纳

上海项目朱家角水上宾馆

工程地点：上海朱家角
结构形式：混凝土框架结构
用地面积：4.30万平方米
建筑面积：4万平方米
设计时间：2007年
竣工时间：2013年

合作建筑师：单立欣、刘恒

保定市博物馆/大剧院

工程地点：河北省保定市
结构形式：混凝土框架-剪力墙结构，屋盖采用钢结构
用地面积：7.05万平方米
建筑面积：6.67万平方米
设计时间：2011年

合作建筑师：单立欣、李斌、董元铮、叶水清

徐州建筑职业技术学院图书馆

工程地点：江苏省徐州市
结构形式：混凝土框架结构
用地面积：5.49万平方米
建筑面积：2.79万平方米
设计时间：2009年
竣工时间：2013年

合作建筑师：赵晓刚、周力坦

昆山博物馆

工程地点：江苏省昆山市
结构形式：混凝土框架-剪力墙结构
用地面积：0.62万平方米
建筑面积：1.07万平方米
设计时间：2011年

合作建筑师：关飞、曾瑞

谷泉会议中心

工程地点：北京市平谷区
结构形式：混凝土框架-剪力墙及局部钢结构
用地面积：25.10万平方米
建筑面积：4.45万平方米
设计时间：2010年
竣工时间：2012年

合作建筑师：周旭梁、时红

青海玉树康巴艺术中心

工程地点：青海省玉树州
结构形式：混凝土框架-剪力墙结构
用地面积：2.45万平方米
建筑面积：2.06万平方米
设计时间：2011年

合作建筑师：关飞、曾瑞、康凯、吴健

北京奥林匹克公园瞭望塔

工程地点：北京奥林匹克公园
结构形式：塔座大厅为钢筋混凝土框架、剪力墙结构，屋盖为大跨
　　　　　度钢筋混凝土根部加腋梁结构，塔身和塔楼为钢结构
用地面积：0.78万平方米
建筑面积：1.87万平方米
设计时间：2011年
竣工时间：2012年

合作建筑师：康凯、叶水清、吴建

敦煌莫高窟旅游中心

工程地点：甘肃省敦煌市
结构形式：框架剪力墙结构
用地面积：4万平方米
建筑面积：1.04万平方米
设计时间：2009年
竣工时间：2013年

合作建筑师：吴斌

崔愷简介

崔愷,1957年生于北京,1984年毕业于天津大学建筑系,获硕士学位。现任中国建筑设计研究院(集团)副院长、总建筑师,中国工程院院士,国家勘察设计大师,崔愷建筑工作室主持人。曾获得"全国优秀科技工程者"(1997)、"国务院特殊津贴专家"(1998)、"国家人事部有突出贡献的中青年专家"(1999)、"国家百、千、万人工程"人选(1999)、"法国文学与艺术骑士勋章"(2003)、"梁思成建筑奖"(2007)等荣誉;获得国家优秀工程设计金奖1项,银奖8项,铜奖5项;建设部优秀工程设计奖一等奖3项,二等奖14项,三等奖7项。获得亚洲建协金奖1项,国际铜业学会一等奖1项,三等奖1项等专业设计奖项;受邀参加巴黎"中国建筑展"(1996)、意大利"威尼斯双年展第八届国际建筑展"(2003)、台北"城市谣言-华人建筑展"(2004)、首届深圳建筑双年展(2005)、伦敦"创意中国"展(2008)、巴黎"中国当代建筑展"(2008)、纽约"中国本土建筑展"(2008)、布鲁塞尔"心造-中国当代建筑前沿展"(2009)、烟台-成都"另一个、同一个-中国当代建筑的断面"建筑作品展(2010)、北京"重生"-汶川震后重建作品展(2010)、香港"2011-12香港深圳城\建筑双城双年展",并在天津及深圳举办"本土设计-崔愷建筑作品展"(2010);出版学术著作《工程报告》(2002)、《德胜尚城》(2009)、《本土设计》(2009)、《中间建筑》(2010)。

崔愷在中国建筑设计研究院主持的建筑工作室有30多名建筑师跟他一起工作。他在28年的职业生涯中取得了很好的声誉。他的主要作品是关于文化设施例如博物馆、剧院、图书馆、学校、历史建筑更新等,也涉及酒店、办公楼等公共建筑。大部分的作品赢得了地方、国家及国际建筑奖。

Profile

Mr. Cui Kai was born in Beijing in 1957, and graduated from Department of Architecture, Tianjin University in 1984 as a master. Now he is Vice President, Chief Architect of the China Architecture Design & Research Group, Academician of Chinese Academy of Engineering, National Design Master and Principal of Cuikai Architecture Studio. He is also a bearer of numerous prizes, such as National Best Science and Technology Worker(1997), Expert who enjoys special subsidy from the State Council(1998), Excellent Mid-young aged Expert of the Ministry of Personnel(1999), Candidate of "National Hundred, Thousand & Ten Thousand Reserves Project"(1999), French Culture & Art Cavalier Medal(2003),and Liang Sicheng Award(2007), together with 1 Gold Medal ,8 Silver Medals and 5 Bronze medals of National Best Engineering Design, 3 First Prizes, 14 Second Prizes and 7 Third Prizes of the Best Design of the Ministry of Construction; ARCASIA Gold Award,First Prize and Third Prize of the International Copper Association and so on. He was invited to a lot of exhibitions, such as the Chinese Architecture Exhibition in PARIS in 1996, La Biennale di Venezia and the 8th International Architecture Exhibition in Italy in 2003, Urban Rumor-Chinese Architecture Exhibition in Taipei in 2004, the first Shenzhen Architectural Biennale in 2005, Creative China Exhibition in London in 2008, Modern Chinese Architecture Exhibition in Paris in 2008, Native Chinese Architecture Exhibition in New York in 2008, Heart-made: The Cutting-Edge of Chinese Contemporary Architecture in Brussels in 2009, The Same one, Another One–Section of Modern Chinese Architecture: Architecture Exhibition in Yantai and Chengdu in 2010, Reborn-Exhibition of Wenchuan Post Earthquake Reconstruction in Beijing in 2010, 2011-12 Hong Kong and Shenzhen City/Architecture Biennale in Hong Kong and so on. He had his personal exhibition in Tianjin and Shenzhen in 2010, the Native Design, Cui Kai's work: Architecture Exhibition. Mr. Cui Kai also carried some works through the press, such as *Construction Report in 2002, Desheng Up-town in 2006, Native Design in 2009 and Inside-Out in 2010.*

Mr.Cui Kai is also in charge of a architect studio within CAG. There are 30 architects working together . He has 28years professional experience and got very good reputation in China. His works are mainly dialing about cultural facilities such as museum ,theater, library,school, renovation of historical building,as well as hotel and office building .Most of his works win reward of local , state and international level.

崔愷工作室成员 / Cuikai Studio team

王庆国 Wang Qingguo、关飞 Guan Fei、王可尧 Wang Keyao、崔愷 Cui Kai、李斌 Li Bin、吴健 Wu Jian、曹洋 Cao Yang、刘恒 Liu Heng、曾瑞 Zeng Rui、傅晓铭 Fu Xiaoming、高凡 Gao Fan、单立欣 Shan Lixin、梁云姣 Liang Yunjiao、朱巍 Zhu Wei、单晓宇 Shan Xiaoyu、李喆 Li Zhe、康凯 Kang Kai、石磊 Shi Lei、张男 Zhang Nan、金爽铮 Jin Shuang、董元正 Dong Yuanzheng、梁丰 Liang Feng、叶水清 Ye Shuiqing、张英奇 Zhang Yingqi、彭彦 Peng Yan、喻弢 Yu Tao、时红 Shi Hong、陈梦津(法国)Chen Aurelien(France)、刘洋 Liu Yang、宋建军 Song Jianjun、周力坦 Zhou Litan、任祖华 Ren Zhuhua、郭海鞍 Guo Haian、马欣 Ma Xin、邢野 Xing Ye、冯君 Feng Jun

曾经成员 / Former collaborators

潘观爱(中国澳门)、郑萌、熊明倩、Eric Daniel Spencer(美国)、Hennecke Christian(德国)、周旭梁、邓烨、吴斌、赵晓刚、张汝冰、李斌、张军英
Pan Guanai(China Macao),Zheng Meng,Xiong Mingqian,Eric Daniel Spencer(USA),Hennecke Christian(Germany),Zhou Xuliang,Deng Ye,Wu Bin,Zhao Xiaogang,Zhang Rubing,Li Bin,Zhang Junying